Lecture Notes in Physics

Editorial Board

R. Beig, Wien, Austria
J. Ehlers, Potsdam, Germany
U. Frisch, Nice, France
K. Hepp, Zürich, Switzerland
W. Hillebrandt, Garching, Germany
D. Imboden, Zürich, Switzerland
R. L. Jaffe, Cambridge, MA, USA
R. Kippenhahn, Göttingen, Germany
R. Lipowsky, Golm, Germany
H. v. Löhneysen, Karlsruhe, Germany
I. Ojima, Kyoto, Japan
H. A. Weidenmüller, Heidelberg, Germany
J. Wess, München, Germany
J. Zittartz, Köln, Germany

Springer
Berlin
Heidelberg
New York
Barcelona
Hong Kong
London
Milan
Paris
Singapore
Tokyo

Physics and Astronomy ONLINE LIBRARY
http://www.springer.de/phys/

The Editorial Policy for Proceedings

The series Lecture Notes in Physics reports new developments in physical research and teaching – quickly, informally, and at a high level. The proceedings to be considered for publication in this series should be limited to only a few areas of research, and these should be closely related to each other. The contributions should be of a high standard and should avoid lengthy redraftings of papers already published or about to be published elsewhere. As a whole, the proceedings should aim for a balanced presentation of the theme of the conference including a description of the techniques used and enough motivation for a broad readership. It should not be assumed that the published proceedings must reflect the conference in its entirety. (A listing or abstracts of papers presented at the meeting but not included in the proceedings could be added as an appendix.)
When applying for publication in the series Lecture Notes in Physics the volume's editor(s) should submit sufficient material to enable the series editors and their referees to make a fairly accurate evaluation (e.g. a complete list of speakers and titles of papers to be presented and abstracts). If, based on this information, the proceedings are (tentatively) accepted, the volume's editor(s), whose name(s) will appear on the title pages, should select the papers suitable for publication and have them refereed (as for a journal) when appropriate. As a rule discussions will not be accepted. The series editors and Springer-Verlag will normally not interfere with the detailed editing except in fairly obvious cases or on technical matters.
Final acceptance is expressed by the series editor in charge, in consultation with Springer-Verlag only after receiving the complete manuscript. It might help to send a copy of the authors' manuscripts in advance to the editor in charge to discuss possible revisions with him. As a general rule, the series editor will confirm his tentative acceptance if the final manuscript corresponds to the original concept discussed, if the quality of the contribution meets the requirements of the series, and if the final size of the manuscript does not greatly exceed the number of pages originally agreed upon. The manuscript should be forwarded to Springer-Verlag shortly after the meeting. In cases of extreme delay (more than six months after the conference) the series editors will check once more the timeliness of the papers. Therefore, the volume's editor(s) should establish strict deadlines, or collect the articles during the conference and have them revised on the spot. If a delay is unavoidable, one should encourage the authors to update their contributions if appropriate. The editors of proceedings are strongly advised to inform contributors about these points at an early stage.
The final manuscript should contain a table of contents and an informative introduction accessible also to readers not particularly familiar with the topic of the conference. The contributions should be in English. The volume's editor(s) should check the contributions for the correct use of language. At Springer-Verlag only the prefaces will be checked by a copy-editor for language and style. Grave linguistic or technical shortcomings may lead to the rejection of contributions by the series editors. A conference report should not exceed a total of 500 pages. Keeping the size within this bound should be achieved by a stricter selection of articles and not by imposing an upper limit to the length of the individual papers. Editors receive jointly 30 complimentary copies of their book. They are entitled to purchase further copies of their book at a reduced rate. As a rule no reprints of individual contributions can be supplied. No royalty is paid on Lecture Notes in Physics volumes. Commitment to publish is made by letter of interest rather than by signing a formal contract. Springer-Verlag secures the copyright for each volume.

The Production Process

The books are hardbound, and the publisher will select quality paper appropriate to the needs of the author(s). Publication time is about ten weeks. More than twenty years of experience guarantee authors the best possible service. To reach the goal of rapid publication at a low price the technique of photographic reproduction from a camera-ready manuscript was chosen. This process shifts the main responsibility for the technical quality considerably from the publisher to the authors. We therefore urge all authors and editors of proceedings to observe very carefully the essentials for the preparation of camera-ready manuscripts, which we will supply on request. This applies especially to the quality of figures and halftones submitted for publication. In addition, it might be useful to look at some of the volumes already published. As a special service, we offer free of charge LaTeX and TeX macro packages to format the text according to Springer-Verlag's quality requirements. We strongly recommend that you make use of this offer, since the result will be a book of considerably improved technical quality. To avoid mistakes and time-consuming correspondence during the production period the conference editors should request special instructions from the publisher well before the beginning of the conference. Manuscripts not meeting the technical standard of the series will have to be returned for improvement.
For further information please contact Springer-Verlag, Physics Editorial Department II, Tiergartenstrasse 17, D-69121 Heidelberg, Germany

Series homepage – http://www.springer.de/phys/books/lnpp

Heinz-Peter Breuer Francesco Petruccione (Eds.)

Relativistic Quantum Measurement and Decoherence

Lectures of a Workshop
Held at the Istituto Italiano per gli Studi Filosofici
Naples, April 9–10, 1999

Springer

Editors

Heinz-Peter Breuer
Francesco Petruccione
University of Freiburg
Faculty of Physics
Hermann-Herder-Strasse 3
79104 Freiburg, Germany

Library of Congress Cataloging-in-Publication Data applied for.

Die Deutsche Bibliothek - CIP-Einheitsaufnahme

Relativistic quantum measurement and decoherence : lectures of a workshop held at the Istituto Italiano per gli Studi Filosofici, Naples, April 9 - 11, 1999 / Heinz-Peter Breuer ; Francesco Petruccione (ed.). - Berlin ; Heidelberg ; New York ; Barcelona ; Hong Kong ; London ; Milan ; Paris ; Singapore ; Tokyo : Springer, 2000
(Lecture notes in physics ; Vol. 559)
(Physics and astronomy online library)
ISBN 3-540-41061-9

ISSN 0075-8450
ISBN 3-540-41061-9 Springer-Verlag Berlin Heidelberg New York

Springer-Verlag Berlin Heidelberg New York
a member of BertelsmannSpringer Science+Business Media GmbH

Typesetting: Camera-ready by the authors/editor
Cover design: *design & production*, Heidelberg

Printed on acid-free paper
SPIN: 10783595 57/3141/du - 5 4 3 2 1 0

Preface

The development of a consistent picture of the processes of decoherence and quantum measurement is among the most interesting fundamental problems with far-reaching consequences for our understanding of the physical world. A satisfactory solution of this problem requires a treatment which is compatible with the theory of relativity, and many diverse approaches to solve or circumvent the arising difficulties have been suggested. This volume collects the contributions of a workshop on *Relativistic Quantum Measurement and Decoherence* held at the Istituto Italiano per gli Studi Filosofici in Naples, April 9-10, 1999. The workshop was intended to continue a previous meeting entitled *Open Systems and Measurement in Relativistic Quantum Theory*, the talks of which are also published in the Lecture Notes in Physics Series (Vol. 526).

The different attitudes and concepts used to approach the decoherence and quantum measurement problem led to lively discussions during the workshop and are reflected in the diversity of the contributions. In the first article the measurement problem is introduced and the various levels of compatibility with special relativity are critically reviewed. In other contributions the rôles of non-locality and entanglement in quantum measurement and state vector preparation are discussed from a pragmatic quantum-optical and quantum-information perspective. In a further article the viewpoint of the consistent histories approach is presented and a new criterion is proposed which refines the notion of consistency. Also, the phenomenon of decoherence is examined from an open system's point of view and on the basis of superselection rules employing group theoretic and algebraic methods. The notions of hard and soft superselection rules are addressed, as well as the distinction between real and apparent loss of quantum coherence. Furthermore, the emergence of real decoherence in quantum electrodynamics is studied through an investigation of the reduced dynamics of the matter variables and is traced back to the emission of bremsstrahlung.

It is a pleasure to thank Avv. Gerardo Marotta, the President of the Istituto Italiano per gli Studi Filosofici, for suggesting and making possible an interesting workshop in the fascinating environment of Palazzo Serra di Cassano. Furthermore, we would like to express our gratidude to Prof. Antonio Gargano, the General Secretaty of the Istituto Italiano per gli Studi Filosofici, for his friendly and efficient local organization. We would also like to thank the participants of the workshop.

Freiburg im Breisgau,
July 2000

Heinz-Peter Breuer
Francesco Petruccione

Contents

List of Participants

Albert, David Z.
Department of Philosophy
Columbia University
1150 Amsterdam Avenue
New York, NY 10027, USA
da5@columbia.edu

Braunstein, Samuel L.
SEECS, Dean Street
University of Wales
Bangor LL57 1UT, United Kingdom
schmuel@sees.bangor.ac.uk

Breuer, Heinz-Peter
Fakultät für Physik
Universität Freiburg
Hermann-Herder-Str. 3
D-79104 Freiburg i. Br., Germany
breuer@physik.uni-freiburg.de

Giulini, Domenico
Universität Zürich
Insitut für Theoretische Physik
Winterthurerstr. 190
CH-8057 Zürich, Schweiz
giulini@physik.unizh.ch

Kent, Adrian
Department of Applied mathematics and Theoretical Physics
University of Cambridge
Silver Street
Cambridge CB3 9EW, United Kingdom
A.P.A.Kent@damtp.cam.ac.uk

Petruccione, Francesco
Fakultät für Physik
Universität Freiburg
Hermann-Herder-Str. 3
D-79104 Freiburg i. Br., Germany
and

Istituto Italiano per gli Studi Filosofici
Palazzo Serra di Cassano
Via Monte di Dio, 14
I-80132 Napoli, Italy
petruccione@physik.uni-freiburg.de

Popescu, Sandu
H. H. Wills Physics Laboratory
University of Bristol
Tyndall Avenue
Bristol BS8 1TL, United Kingdom
and
BRIMS, Hewlett-Packard Laboratories
Stoke Gifford
Bristol, BS12 6QZ, United Kingdom
S.Popescu@bris.ac.uk

Unruh, William G.
Department of Physics
University of British Columbia
6224 Agricultural Rd.
Vancouver, B. C., Canada V6T1Z1
unruh@physics.ubc.ca

Special Relativity as an Open Question

David Z. Albert

Department of Philosophy, Columbia University, New York, USA

Abstract. There seems to me to be a way of reading some of the trouble we have lately been having with the quantum-mechanical measurement problem (not the *standard* way, mind you, and certainly not the *only* way; but a way that nonetheless be worth exploring) that suggests that there are fairly prosaic physical circumstances under which it might not be entirely beside the point to look around for *observable violations* of *the special theory of relativity*. The suggestion I have in mind is connected with attempts over the past several years to write down a relativistic field-theoretic version of the dynamical reduction theory of Ghirardi, Rimini, and Weber [Physical Review **D34**, 470-491 (1986)], or rather it is connected with the persistent *failure* of those attempts, it is connected with the most obvious strategy for giving those attempts *up*. And that (in the end) is what this paper is going to be about.

1 The Measurement Problem

Let me start out (however) by reminding you of precisely what the quantum-mechanical problem of measurement *is*, and then talk a bit about where things stand at present vis-a-vis the general question of the compatibility of quantum mechanics with the special theory of relativity, and then I want to present the simple, standard, well-understood non-relativistic version of the Ghirardi, Rimini, and Weber (GRW) theory [1], and *then* (at last) I will get into the business I referred to above.

First the measurement problem. Suppose that every system in the world invariably evolves in accordance with the linear deterministic quantum-mechanical equations of motion and suppose that M is a good measuring instrument for a certain observable A of a certain physical system S. What it means for M to be a "good" measuring instrument for A is just that for all eigenvalues a_i of A:

$$|\text{ready}\rangle_M |A = a_i\rangle_S \longrightarrow |\text{indicates that } A = a_i\rangle_M |A = a_i\rangle_S, \tag{1}$$

where $|\text{ready}\rangle_M$ is that state of the measuring instrument M in which M is prepared to carry out a measurement of A, "$\longrightarrow$" denotes the evolution of the state of $M + S$ during the measurement-interaction between those two systems, and $|\text{indicates that } A = a_i\rangle_M$ is that state of the measuring instrument in which, say, its pointer is pointing to the the a_i-position on its dial. That is: what it means for M to be a "good" measuring instrument for A is just that M invariably indicates the correct value for A in all those states of S in which A *has* any definite value.

The problem is that (1), together with the linearity of the equations of motion entails that:

$$\sum_i |\text{ready}\rangle_M |A = a_i\rangle_S \longrightarrow \sum_i |\text{indicates that } A = a_i\rangle_M |A = a_i\rangle_S. \qquad (2)$$

And that appears not to be what actually happens in the world. The right-hand side of Eq. (2) is (after all) a *superposition* of various different outcomes of the A-measurement - and decidedly *not* any particular *one* of them. But what actually *happens* when we measure A on a system S in a state like the one on the left-hand-side of (2) is of course that *one* or *another* of those particular outcomes, and nothing else, *emerges.*

And there are two big ideas about what to *do* about that problem that seem to me to have any chance at all of being on the right track.

One is to deny that the standard way of thinking about what it means to be in a superposition is (as a matter of fact) the right way of thinking about it; to deny, for example, that there fails to be any determinate matter of fact, when a quantum state like the one here obtains, about where the pointer is pointing.

The idea (to come at it from a slightly different angle) is to construe quantum-mechanical wave-functions as *less than complete descriptions of the world.* The idea that something *extra* needs to be *added* to the wave-function description, something that can broadly be thought of as *choosing between* the two conditions superposed here, something that can be thought of as somehow *marking* one of those two conditions as the unique, *actual*, outcome of the measurement that leads up to it.

Bohm's theory is a version of this idea, and so are the various modal interpretations of quantum mechanics, and so (more or less) are many-minds interpretations of quantum mechanics.[1]

The other idea is to stick with the standard way of thinking about what it means to be in a superposition, and to stick with the idea that a quantum-mechanical wave-function amounts, all by itself, to a complete description of a physical system, and to account for the emergence of determinate outcomes of experiments like the one we were talking about before by means of explicit *violations* of the linear deterministic equations of motion, and to try to develop some precise idea of the circumstances under which those violations occur.

And there is an enormously long and mostly pointless history of speculations in the physical literature (speculations which have notoriously hinged on distinctions between the "microscopic" and the "macroscopic", or between

[1] Many-minds interpretations are a bit of a special case, however. The outcomes of experiments on those interpretations (although they are perfectly actual) are not unique. The more important point, though, is that those interpretations (like the others I have just mentioned) solve the measurement problem by construing wave-functions as incomplete descriptions of the world.

the "reversible" and the "irreversible", or between the "animate" and the "inanimate", or between "subject" and "object", or between what does and what doesn't genuinely amount to a "measurement") about precisely what sorts of violations of those equations - what sorts of *collapses* - are called for here; but there has been to date only one fully-worked-out, traditionally scientific sort of proposal along these lines, which is the one I mentioned at the beginning of this paper, the one which was originally discovered by Ghirardi and Rimini and Weber, and which has been developed somewhat further by Philip Pearle and John Bell.

There are (of course) other traditions of thinking about the measurement problem too. There is the so-called Copenhagen interpretation of quantum mechanics, which I shall simply leave aside here, as it does not even pretend to amount to a realistic description of the world. And there is the tradition that comes from the work of the late Hugh Everett, the so called "many worlds" tradition, which is (at first) a thrilling attempt to have one's cake and eat it too, and which (more particularly) is committed *both* to the proposition that quantum-mechanical wave-functions are complete descriptions of physical systems *and* to the proposition that those wave-functions invariably evolve in accord with the standard linear quantum-mechanical equations of motion, and which (alas, for a whole bunch of reasons) seems to me not to be a particular candidate either.[2]

And that's about it.

2 Degrees of Compatibility with Special Relativity

Now, the story of the compatibility of these attempts at solving the measurement problem with the special theory of relativity turns out to be unexpectedly rich. It turns out (more particularly) that compatibility with special relativity is the sort of thing that admits of *degrees*. We will need (as a matter of fact) to think about *five* of them - not (mind you) because only five are logically imaginable, but because one or another of those five corresponds to every one of the fundamental physical theories that anybody has thus far taken seriously.

Let's start at the top.

What it is for a theory to be *metaphysically* compatible with special relativity (which is to say: what it is for a theory to be compatible with special relativity in the *highest* degree) is for it to depict the world as unfolding in a four-dimensional Minkowskian space-time. And what it means to speak of the world as unfolding within a four-dimensional Minkovskian space-time is (i) that everything there is to say about the world can straightforwardly be

[2] Foremost among these reasons is that the many-worlds interpretations seems to me not to be able to account for the facts about chance. But that's a long story, and one that's been told often enough elsewhere.

read off of a catalogue of the *local physical properties* at every one of the continuous infinity of *positions* in a space-time like that, and (ii) that whatever lawlike *relations* there may be between the values of those local properties can be *written down* entirely in the *language* of a space-time that - that whatever lawlike relations there may be between the values of those local properties are *invariant* under *Lorentz-transformations.* And what it is to pick out some particular *inertial frame of reference* in the context of the sort of theory we're talking about here - what it is (that is) to *adopt the conventions of measurement* that are indigenous to any particular frame of reference in the context of the sort of theory we're talking about here - is just to pick out some particular way of organising everything there is to say about the world into a *story*, into a *narrative*, into a *temporal sequence* of *instantaneous global physical situations.* And every possible world on such a theory will invariably be organizable into an *infinity* of such stories - and those stories will invariably be related to one another by Lorentz-transformations. And note that if even a single one of those stories is in accord with the *laws*, then (since the laws are *invariant* under Lorentz-transformations) *all* of them must be.

The Lorentz-invariant theories of classical physics (the electrodynamics of Maxwell, for example) are metaphysically compatible with special relativity; and so (more surprisingly) are a number of radically *non-local* theories (completely hypothetical ones, mind you - ones which in so far as we know at present have no application whatever to the actual world) which have recently appeared in the literature.[3]

But it happens that not a single one of the existing proposals for making sense of *quantum mechanics* is metaphysically compatible with special relativity, and (moreover) it isn't easy to imagine there ever *being* one that *is.* The reason is simple: What is absolutely *of the essence* of the quantum-mechanical picture of the world (in so far as we understand it at present), what none of the attempts to straighten quantum mechanics out have yet dreamed of dispensing with, are *wave-functions.* And wave-functions just don't *live* in four-dimensional space-times; wave-functions (that is) are just *not* the sort of objects which can always be uniquely picked out by means of any catalogue of the local properties of the positions of a space-time like that. As a general matter, they need *bigger* ones, which is to say *higher-dimensional* ones, which is to say *configurational* ones. And that (alas!) is that.

The next level down (let's call this one the level of *dynamical* compatibility with special relativity) is inhabited by pictures on which the physics of the world is exhaustively described by something along the lines of a (so-called) relativistic quantum field theory - a *pure* one (mind you) in which

[3] Tim Maudlin and Frank Artzenius have both been particularly ingenious in concocting theories like these, which (notwithstanding their non-locality) are entirely formulable in four-dimensional Minkowski space-time. Maudlin's book *Quantum Non-Locality and Relativity* (Blackwell, 1994) contains extremely elegant discussions of several such theories.

there are no additional variables, and in which the quantum states of the world invariably evolve in accord with local, deterministic, Lorentz-invariant quantum mechanical equations of motion. These pictures (once again) must depict the world as unfolding not in a *Minkowskian* space-time but in a *configuration* one - and the *dimensionalities* of the configuration space-times in question here are (of course) going to be infinite. Other than that, however, everything remains more or less as it was above. The configuration space-time in question here is built directly *out* of the Minkowskian one (remember) by treating each of the points in Minkowskian space-time (just as one does in the classical theory of fields) as an instantaneous bundle of *physical degrees of freedom*. And so what it is to pick out some particular *inertial frame of reference* in the context of this sort of picture is still just to pick out some particular way of organizing everything there is to say about the world into a temporal sequence of instantaneous global physical situations. And every possible world on this sort of a theory will still be organizable into an *infinity* of such stories. And those stories will still be related to one another by means of the appropriate generalizations of the Lorentz point-transformations. And it will still be the case that if even a single one of those stories is in accord with the laws, then (since the laws are invariant under Lorentz-transformations) all of them must be.

The trouble is that there may well not *be* any such pictures that turn out to be worth taking seriously. All we have along these lines at present (remember) are the many-worlds pictures (which I fear will turn out not to be coherent) and the many-*minds* pictures (which I fear will turn out not to be *plausible*).

And further down things start to get ugly.

We have known for more than thirty years now that any proposal for making sense of quantum mechanics on which measurements invariably have unique and particular and determinate *outcomes* (which covers all of the proposals I know about, or at any rate the ones I know about that are also worth *thinking* about, other than many worlds and many minds) is going to have no choice whatever but to turn out to be *non-local*.

Now, non-locality is certainly not an obstacle *in and of itself* even to *metaphysical* compatibility with special relativity. There are now (as I mentioned before) a *number* of explicit examples in the literature of hypothetical dynamical laws which are radically non-local and which are nonetheless cleverly cooked up in such a way as to be formulable entirely within Minkowski-space. The thing is that none of them can even remotely mimic the empirical predictions of *quantum mechanics*; and that nobody I talk to thinks that we have even the slightest reason to hope for one that will.

What we *do* have (on the other hand) is a very straightforward trick by means of which a wide variety of theories which are radically non-local and (moreover) are flatly incompatible with the proposition that the stage on which physical history unfolds is Minkowki-space can nonetheless be made fully and trivially *Lorentz-invariant*; a trick (that is) by means of which a wide variety of such theories can be made what you might call *formally* compatible with special relativity.

The trick [2] is just to let go of the requirement that the physical history of the world can be represented in its entirety as a temporal sequence of *situations*. The trick (more particularly) is to let go of the requirement that the situation associated with two intersecting space-like hypersurfaces in the Minkowski-space must agree with one another about the expectations values of local observables at points where the two surfaces coincide.

Consider (for example) an old-fashioned non-relativistic projection-postulate, which stipulates that the quantum states of physical systems invariably evolve in accord with the linear deterministic equations of motion *except* when the system in question is being "measured"; and that the quantum state of a system instantaneously *jumps*, at the instant the system *is* measured, into the eigenstate of the measured observable corresponding to the outcome of the measurement. This is the sort of theory that (as I mentioned above) nobody takes seriously anymore, but never mind that; it will serve us well enough, for the moment, as an illustration. Here's how to make this sort of a projection-postulate Lorentz-invariant: First, take the linear collapse-free dynamics of the measured system - the dynamics which we are generally in the habit of writing down as a deterministic connection between the wave-functions on two arbitrary equal-time-hyperplanes - and *re*-write it as a deterministic connection between the wave-functions on two arbitrary *space-like-hypersurfaces*, as in Fig. 1. Then stipulate that the jumps referred to above occur not (as it were) when the equal-time-hyperplane sweeps across the measurement-event, but whenever an arbitrary space-like hypersurface *undulates* across it.

Suppose (say) that the momentum of a free particle is measured along the hypersurface marked $t = 0$ in Fig. 2, and that later on a measurement locates the particle at P. Then our new projection-postulate will stipulate among other things) that the wave-function of the particle along hypersurface a is an eigenstate of momentum, and that the wave-function of the particle along hypersurface b is (very nearly) an eigenstate of *position*. And *none* of that (and nothing *else* that this new postulate will have to say) refers in any way shape or form to any particular *Lorentz frame*. And this is pretty.

But think for a minute about what's been *paid* for it. As things stand now we have let go not only of *Minkowski-space* as a realistic description of the stage on which the story of the world is enacted, but (in so far as I can see) of *any* conception of that stage *whatever*. As things stand now (that is) we have let go of the idea of the world's having anything along the lines of a narratable story *at all*! And all this just so as to guarantee that the fundamental laws remain exactly invariant under a certain hollowed-out set of *mathematical transformations*, a set which is now of no particular deep conceptual *interest*, a set which is now utterly disconnected from any idea of an *arena* in which the world occurs.

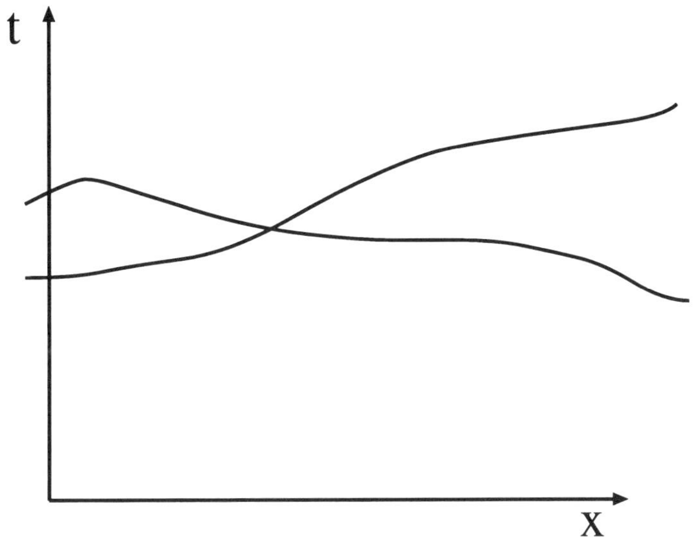

Fig. 1. Two arbitrary spacelike hypersurfaces.

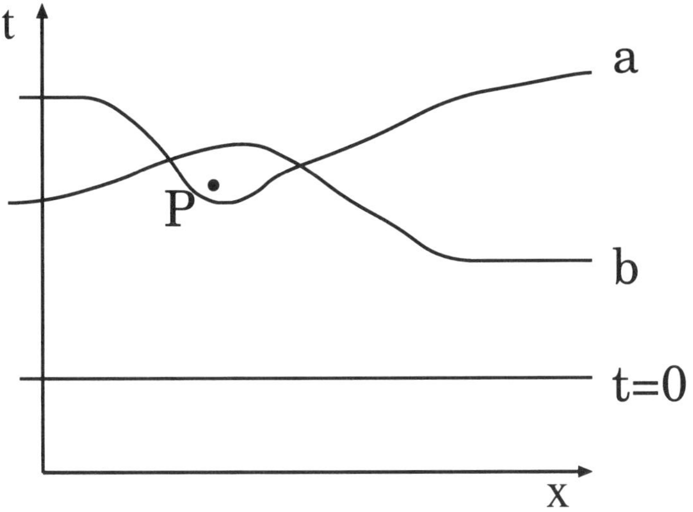

Fig. 2. A measurement locates a free particle in P (see text).

Never mind. Suppose we had somehow managed to *resign* ourselves to that. There would *still* be trouble. It happens (you see) that notwithstanding the enormous energy and technical ingenuity has been expended over the past several years in attempting to concoct a version of a more *believable* theory of collapses - a version (say) of *GRW* theory - on which a trick like this might work, even *that* (paltry as it is) is as yet beyond our grasp.

And all that (it seems to me) ought to give us pause.

The next level down (let's call this one the level of discreet *in*compatibility with special relativity) is inhabited by theories (Bohm's theory, say, or modal theories) on which the special theory of relativity, *whatever* it means, is unambiguously *false*; theories (that is) which explicitly *violate* Lorentz-invariance, but which nonetheless manage to *refrain* from violating it in any of their predictions about the *outcomes* of *experiments*. These theories (to put it slightly differently) all require that there be some legally privileged Lorentz-frame, but they all also entail that (as a matter of fundamental principle) no performable experiment can identify *what frame that is*.

And then (at last) there are theories that explicitly violate Lorentz-invariance (but presumably only a bit, or only in places we haven't looked yet) even in their observable predictions. It's *that* sort of a theory (the sort of a theory we'll refer to as *manifestly* incompatible with special relativity) that I'm going to want to draw your attention to here.

3 The Theory I Have in Mind

But one more thing needs doing before we get to that, which is to say something about where the theory I have in mind *comes* from. And where it comes from (as I mentioned at the outset) is the non-relativistic spontaneous localization theory of Ghirardi, Rimini, Weber, and Pearle.

GRWP's idea was that the wave function of an N-particle system

$$\psi(\boldsymbol{r}_1, \boldsymbol{r}_2, \ldots, \boldsymbol{r}_N, t) \tag{3}$$

usually evolves in the familiar way - in accordance with the Schrödinger equation - but that every now and then (once in something like $10^{15}/N$ seconds), at random, but with fixed probability per unit time, the wave function is suddenly multiplied by a normalized Gaussian (and the product of those two *separately* normalized functions is multiplied, at that same instant, by an overall renormalizing constant). The form of the multiplying Gaussian is:

$$K \exp\left[-(\boldsymbol{r} - \boldsymbol{r}_k)^2/2\sigma^2\right] \tag{4}$$

where $\boldsymbol{r}_k$ is chosen at random from the arguments $\boldsymbol{r}_n$, and the width σ of the Gaussian is of the order of 10^{-5} cm. The *probability* of this Gaussian being centered at any particular point $\boldsymbol{r}$ is stipulated to be proportional to the absolute square of the inner product of (3) (evaluated at the instant just prior to this "jump") with (4). Then, until the next such "jump", everything proceeds as before, in accordance with the Schrödinger equation. The probability of such jumps per particle per second (which is taken to be something like 10^{-15}, as I mentioned above), and the width of the multiplying Gaussians (which is taken to be something like 10^{-5} cm) are new constants of nature.

That's the whole theory. No attempt is made to explain the occurrence of these "jumps"; that such jumps occur, and occur in precisely the way stipulated above, can be thought of as a new fundamental law; a beautiful and absolutely explicit *law of collapse*, wherein there is no talk at a fundamental level of "measurements" or "recordings" or "macroscopicness" or anything like that.

Note that for isolated microscopic systems (i.e. systems consisting of small numbers of particles) "jumps" will be rare as to be completely unobservable in practice; and the width of the multiplying Gaussian has been chosen large enough so that the violations of conservation of energy which those jumps will necessarily produce will be very small (over reasonable time-intervals), even for macroscopic systems.

Moreover, if it's the case that every measuring instrument worthy of the name has got to include some kind of a *pointer*, which indicates the outcome of the measurement, and if that pointer has got to be a macroscopic physical object, and if that pointer has got to assume macroscopically different spatial positions in order to indicate different such outcomes (and all of this seems plausible enough, at least at first), then the GRW theory can apparently guarantee that all measurements have *outcomes*. Here's how: Suppose that the GRW theory is true. Then, for measuring instruments (M) such as were just described, superpositions like

$$|A\rangle|M \text{ indicates that } A\rangle + |B\rangle|M \text{ indicates that } B\rangle \tag{5}$$

(which will invariably be superpositions of macroscopically different localized states of some macroscopic physical object) are just the sorts of superpositions that don't last long. In a very short time, in only as long as it takes for the pointer's wave-function to get multiplied by one of the GRW Gaussian (which will be something of the order of $10^{15}/N$ seconds, where N is the number of elementary particles in the pointer) one of the terms in (5) will disappear, and only the other will propagate. Moreover, the probability that one term rather than another survives is (just as standard Quantum Mechanics dictates) proportional to the fraction of the norm which it carries.

And maybe it's worth mentioning here that there are two reasons why this particular way of making experiments have outcomes strikes me at present as conspicuously more interesting than others I know about.

The first has to do with questions of ontological parsimony: We have no way whatever of making experiments have outcomes (after all) that does *without* wave-functions. And only many-worlds theories and collapse theories manage to do without anything *other* than wave-functions[4]. And many-worlds theories don't appear to *work*.

The second (which strikes me as more important) is that the GRW theory affords a means of reducing the probabilities of Statistical Mechanics entirely to the probabilities of Quantum Mechanics. It affords a means (that is) of rearranging the foundations of the entirety of physics so as to contain exactly *one* species of chance. And no other way we presently have of making measurements have outcomes - not Bohm's theory and not modal theories and not many-minds theories and not many-worlds theories and not the Copenhagen interpretation and not quantum logic and not even the other *collapse* theories presently on the market - can do anything *like* that.[5] But let me go back to my story.

4 Approximate Compatibility with Special Relativity

The trouble (as we've seen) is that there can probably not be a version of a theory like this which has any sorts of compatibility with special relativity that seem worth wanting.

And the question is what to *do* about that.

And one of the things it seems to me one might do is to begin to wonder exactly what the all the fuss has been about. One of the things it seems to me one might do - given that the theory of relativity is already *off the table* here as a *realistic description of the structure of the world* - is to begin to wonder exactly what the point is of entertaining only those fundamental theories which are strictly invariant under Lorentz transformations, or even only those fundamental theories whose *empirical predictions* are strictly invariant under Lorentz transformations.

Why not theories which are are only *approximately* so? Why not theories which violate Lorentz invariance in ways which we would be unlikely to have *noticed* yet? Theories like *that*, and (more particularly) GRW-like theories like that, turn out to be *snap* to cook up.

Let's (finally) think one through. Take (say) standard, Lorentz-invariant, relativistic quantum electrodynamics - *without* a collapse. And add to it some non-Lorentz-invariant second-quantized generalization of a collapse-process which is designed to reduce - under appropriate circumstances, *and in some particular preferred frame of reference* - to a standard non-relativistic GRW Gaussian collapse of the effective wave-function of electrons. And suppose

[4] All other pictures (Bohm, Modal Interpretations, Many-Minds, etc) supplement the wave-function with something *else*; something which we know there to *be* a way of doing without; something which (when you think about it this way) looks as if it must somehow be *superfluous*.

[5] This is one of the main topics of a book I have just finished writing, called *Time and Chance*, which is to be published in the fall of 2000 by Harvard University Press.

that the frame associated with our *laboratory* is some frame *other* than the preferred one. And consider what measurements carried out in that laboratory will show.

This needs to be done with some care. What happens in the lab frame is certainly *not* (for example) that the wave-function gets multiplied by anything along the lines of a "*Lorentz-transformation*" of the non-relativistic GRW Gaussian I mentioned a minute ago, for the simple reason that Gaussians are not the sort of things that are *susceptible* of having a Lorentz transformation carried out on them in the *first* place.[6] And it is (as a more general matter) *not to be expected* that a theory like this one is going to yield any straightforward universal geometrical technique whatever - such as we have always had at our disposal, in one form or another, throughout the entire modern history of physics - whereby the way the world looks to one observer can be read off of the way it looks to some *other* one, who is in constant rectilinear motion relative to the first. The theory we have in front of us at the moment is simply not *like* that. We are (it seems fair to say) in infinitely messier waters here. The only absolutely reliable way to proceed on theories like this one (unless and until we can argue otherwise) is to deduce how things may look to this or that observer by explicitly treating those observers and all of their measuring instruments as ordinary physical objects, whose states change only and exactly in whatever way it is that they are required to change by the microscopic laws of nature, and whose evolutions will presumably need to be calculated from the point of view of the unique frame of reference in which those laws take on their simplest form.

That having been said, remember that the violations of Lorentz-invariance in this theory arise *exclusively* in connection with *collapses*, and that the *collapses* in this theory have been specifically designed so as to have no effects whatever, or no effects to *speak* of, on any of the familiar properties or behaviours of everyday localized solid macroscopic objects. And so, in so far as we are concerned with things like (say) the length of medium-sized wooden dowels, or the rates at which cheap spring-driven wristwatches tick, everything is going to proceed, to a very good approximation, as if no such violations were occurring at all.

Let see how far we can run with just that.

Two very schematic ideas for experiments more or less jump right out at you - one of them zeros in on what this theory still has left of the special-relativistic length-contraction, and the other on what it still has left of the special-relativistic time-dilation.

The first would go like this: Suppose that the wave-function of a subatomic particle which is more or less at rest in our lab frame is divided in half - suppose, for example, that the wave-function of a neutron whose z-spin

[6] The sort of thing you need to start out with, if you want to do a Lorentz transformation, is not a function of three-space (which is what a Gaussian is) but a function of three-space and *time*.

is initially "up" is divided, by means of a Stern-Gerlach magnet, into equal y-spin "up" and y-spin "down" components. And suppose that one of those halves is placed in box A and that the other half is placed in box B. And suppose that those two boxes are fastened on to opposite ends of a little wooden dowel. And suppose that they are left in that condition for a certain interval - an interval which is to be measured (by the way) in the *lab* frame, and by means of a co-moving cheap mechanical wristwatch. And suppose that at the end of that interval the two boxes are brought back together and opened, and that we have a look - in the usual way - for the usual sort of interference effects. Note (and this is the crucial point here) that the *length* of this dowel, as measured in the *preferred* frame, will depend radically (if the velocity of the lab frame relative to the preferred one is sufficiently large) on the dowel's *orientation*. If, for example, the dowel is perpendicular to the velocity of the lab relative to the preferred frame, it's length will be the same in the preferred frame as in the lab, but if the the dowel is *parallel* to that relative velocity, then it's length - and hence also the spatial separation between A and B - as measured in the *preferred* frame, will be much *shorter*. And of course the degree to which the GRW collapses *wash out* the interference effects will vary (inversely) with the distance between those boxes as measured in the *preferred* frame.[7] And so it is among the predictions of the sort of theory we are entertaining here that if the lab frame is indeed moving rapidly with respect to the preferred one, the observed interference effects in these sorts of contraptions ought to observably *vary* as the spatial *orientation* of that device is altered. It is among the consequences of the failure of Lorentz-invariance in this theory that (to put it slightly differently) in frames other than the preferred one, invariance under spatial rotations fails as well.

The second experiment involves exactly the same contraption, but in this case what you do with it is to *boost* it - particle, dowel, boxes, wristwatch and all - in various directions, and to various degrees, but always (so as to keep whatever this theory still has in it of the Lorentzian length-contractions entirely out of the picture for the moment) *perpendicular to the length of the dowel.* As viewed in the preferred frame, this will yield interference experiments of different temporal *durations*, in which different numbers of GRW collapses will typically occur, and in which the observed interference effects will (in consequence) be washed out to different *degrees.*

The sizes of these effects are of course going to depend on things like the velocity of the earth relative to the preferred frame (which there can be no

[7] More particularly: If, in the preferred frame, the separation between the two boxes is so small as to be of the order of the width of the GRW Gaussian, the washing-out will more or less vanish altogether.

way of guessing)[8], and the degree to which we are able to boost contraptions of the sort I have been describing, and the accuracies with which we are able to observe interferences, and so on.

The size of the effect in the time-dilation experiment is always going to vary linearly in $\sqrt{1-v^2/c^2}$, where v is the magnitude of whatever boosts we find we are able to artificially produce. In the length-contraction experiment, on the other hand, the effect will tend to pop in and out a good deal more dramatically. If (in that second experiment) the velocity of the contraption relative to the preferred frame can somehow be gotten up to the point at which $\sqrt{1-v^2/c^2}$ is of the order of the width of the GRW Gaussian divided by the length of the dowel - either in virtue of the motion of the earth itself, or by means of whatever boosts we find we are able to artificially produce, or by means of some combination of the two - whatever washing-out there is of the interference effects when the length of the dowel is perpendicular to its velocity relative to the preferred frame will more or less *discontinuously vanish* when we *rotate* it.

Anyway, it seems to me that it might well be worth the trouble to do some of the arithmetic I have been alluding to, and to inquire into some of our present technical capacities, and to see if any of this might actually be worth going out and trying.[9]

References

1. Ghirardi G. C. , Rimini A., Weber T. (1986): *Unified dynamics for microscopic and macroscopic systems.* Phys. Rev. D **34**, 470-491.
2. Aharonov Y., Albert D. (1984): *Is the Familiar Notion of Time-Evolution Adequate for Quantum-Mechanical Systems? Part II: Relativistic Considerations.* Phys. Rev. D **29**, 228-234.

[8] All one can say for certain, I suppose, is that (at the very worst) there must be a time in the course of every terrestrial year at which that velocity is at least of the order of the velocity of the earth relative to the sun.

[9] All of this, of course, leaves aside the question of whether there might be still *simpler* experiments, experiments which might perhaps have already been *performed*, on the basis of which the theory we have been talking about here might be falsified. It goes without saying that I don't (as yet) know of any. But that's not saying much.

Event-Ready Entanglement

Pieter Kok and Samuel L. Braunstein

SEECS, University of Wales, Bangor LL57 1UT, UK

Abstract. We study the creation of polarisation entanglement by means of optical entanglement swapping (Zukowski *et al.*, [*Phys. Rev. Lett.* **71**, 4287 (1993)]). We show that this protocol does not allow the creation of maximal 'event-ready' entanglement. Furthermore, we calculate the outgoing state of the swapping protocol and stress the fundamental physical difference between states in a Hilbert space and in a Fock space. Methods suggested to enhance the entanglement in the outgoing state as given by Braunstein and Kimble [*Nature* **394**, 840 (1998)] generally fail.

1 Introduction

Ever since the seminal paper of Einstein, Podolsky and Rosen [1], the concept of entanglement has captured the imagination of physicists. The EPR paradox, of which entanglement is the core constituent, points out the non-local behaviour of quantum mechanics. This non-locality was quantified by Bell in terms of the so-called Bell inequalities [2] and cannot be explained classically.

Now, at the advent of the quantum information era, entanglement is no longer a mere curiosity of a theory which is highly successful in describing the natural phenomena. It has become an indispensable resource in quantum information protocols such as dense coding, quantum error correction and quantum teleportation [3–6].

Two quantum systems, parametrised by x_1 and x_2 respectively, are called entangled when the state $\Psi(x_1, x_2)$ describing the total system cannot be factorised into states $\psi_1(x_1)$ and $\psi_2(x_2)$ of the separate systems:

$$\Psi(x_1, x_2) \neq \psi_1(x_1)\psi_2(x_2) \, . \tag{1}$$

All the states $\Psi(x_1, x_2)$ accessible to two quantum systems form a set $\mathcal{S}$. These states are generally entangled. Only in extreme cases $\Psi(x_1, x_2)$ is *separable*, i.e., it can be written as a product of states describing the separate systems. The set of separable states form a subset of $\mathcal{S}$ with measure zero.

We arrive at another extremum when the states $\Psi(x_1, x_2)$ are *maximally* entangled. The set of maximally entangled states also forms a subset of $\mathcal{S}$ with measure zero. In order to elaborate on maximal entanglement, we will limit our discussion to quantum optics.

Two photons can be linearly polarised along two orthogonal directions x and y of a given coordinate system. Every possible state of those two photons shared between a pair of modes can be written on the basis of four orthonormal states $|x, x\rangle$, $|x, y\rangle$, $|y, x\rangle$ and $|y, y\rangle$. These basis states generate a four-dimensional Hilbert space. Another possible basis for this space is given by the so-called polarisation Bell states:

$$\begin{aligned} |\Psi^{\pm}\rangle &= (|x,y\rangle \pm |y,x\rangle)/\sqrt{2}\,, \\ |\Phi^{\pm}\rangle &= (|x,x\rangle \pm |y,y\rangle)/\sqrt{2}\,. \end{aligned} \tag{2}$$

These states are also orthonormal. They are examples of *maximally* entangled states. The Bell states are not the only maximally entangled states, but they are the ones most commonly discussed. For the remainder of this paper we will restrict our discussion to the antisymmetric Bell state $|\Psi^-\rangle$ (in the appendix we will explain in more detail why we can do this without loss of generality).

Suppose we want to conduct an experiment which makes use of polarisation entanglement, in particular $|\Psi^-\rangle$. Ideally, we would like to have a source which produces these states at the push of a button. In practice, this might be a bit much to ask. A second option is to have a source which only produces $|\Psi^-\rangle$ randomly, but flashes a red light when it happens. Such a source would create so-called *event-ready* entanglement: it produces $|\Psi^-\rangle$ only part of the time, but when it does, it tells you so.

More formally, the outgoing state $|\psi_{\text{out}|\text{red light flashes}}\rangle$ conditioned on the red light flashing is said to exhibit event-ready entanglement if it can be written as

$$|\psi_{\text{out}|\text{red light flashes}}\rangle \simeq |\Psi^-\rangle + O(\xi)\;, \tag{3}$$

where $\xi \ll 1$. In the remainder of this paper we will omit the subscript '|red light flashes' since it is clear that we can only speak of event-ready entanglement conditioned on the red light's flashing.

Currently, event-ready entanglement has never been produced experimentally. However, non-maximal entanglement has been created by means of, for instance, parametric down-conversion [7]. Rather than a (near) maximally entangled state, as in Eq. (3), this process produces states with a large vacuum contribution. Only a minor part consists of an entangled photon state. Every time parametric down-conversion is employed, there is only a small probability of creating an entangled photon-pair. For the purposes of this paper we will call this *randomly produced* entanglement.

Parametric down-conversion has been used in several experiments, and for some applications randomly produced entanglement therefore seems sufficient. However, on a theoretical level, maximally entangled states appear as primitive notions in many quantum protocols. It is therefore not at all clear whether randomly produced entanglement is suitable for all these cases. This is one of the main motives in our search for event-ready entanglement, where we can ensure that the physical state is maximally entangled.

In this paper we investigate one particular possibility to create event-ready entanglement. It was suggested by Zukowski, Zeilinger, Horne and Ekert [8] and Pavičić [9] that *entanglement swapping* is a suitable candidate. We will therefore study this protocol in some detail using quantum optics.

Entanglement swapping is essentially the teleportation of one part of an entangled pair [3,8,10]. Suppose we have a system of two independent (maximally) entangled photon-pairs in modes a, b and c, d. If we restrict ourselves to the Bell states, we have for instance

$$|\Psi\rangle_{abcd} = |\Psi^-\rangle_{ab} \otimes |\Psi^-\rangle_{cd} \, . \tag{4}$$

However, this state can be written on a different basis:

$$\begin{aligned} |\Psi\rangle_{abcd} = \frac{1}{2}|\Psi^-\rangle_{ad} \otimes |\Psi^-\rangle_{bc} + \frac{1}{2}|\Psi^+\rangle_{ad} \otimes |\Psi^+\rangle_{bc} \\ +\frac{1}{2}|\Phi^-\rangle_{ad} \otimes |\Phi^-\rangle_{bc} + \frac{1}{2}|\Phi^+\rangle_{ad} \otimes |\Phi^+\rangle_{bc} \, . \end{aligned} \tag{5}$$

If we make a Bell measurement on modes b and c, we can see from Eq. (5) that the undetected remaining modes a and d become entangled. For instance, when we find modes b and c in a $|\Phi^+\rangle$ Bell state, the remaining modes a and d must be in the $|\Phi^+\rangle$ state as well.

Although maximally entangled states have never been produced experimentally, entanglement swapping might offer us a solution [9]. An entangled state with a large vacuum contribution (as produced by parametric down-conversion) can only give us randomly produced entanglement. However, if we use two such states and perform entanglement swapping, the Bell detection will act as a tell-tale that there were photons in the system. The question is whether this Bell detection is enough to ensure that an event-ready entangled state appears as a freely propagating wave-function.

Entanglement swapping has been demonstrated experimentally by Pan *et al.* [10], using parametric down-conversion as the entanglement source. In the next section we briefly review the down-conversion mechanism and its role in the experiment of Pan *et al.* Subsequently, in section 3 we study whether entanglement swapping can give us event-ready entanglement.

2 Parametric Down-Conversion and Entanglement Swapping

In this section we review the mechanism of parametric down-conversion and the experimental realisation of entanglement swapping. In parametric down-conversion a crystal is pumped by a high-intensity laser, which we will treat classically (the parametric approximation). The crystal is special in the sense that it has different refractive indices for horizontally and vertically polarised light. In the crystal, a photon from the pump is split into two photons with half the energy of the pump photon. Furthermore, the two photons have orthogonal polarisations. The outgoing modes of the crystal constitute two intersecting cones with orthogonal polarisations as depicted in Fig. 1.

Due to the conservation of momentum, the two produced photons are always in opposite modes with respect to the central axis (determined by

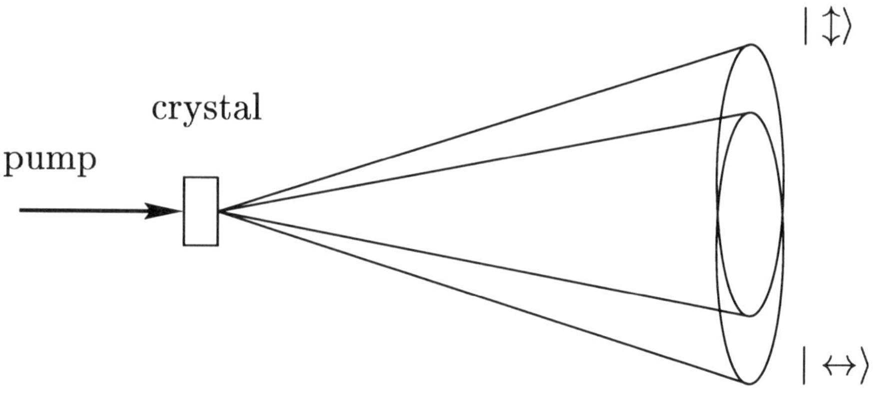

Fig. 1. A schematic representation of type II parametric down-conversion. A high-intensity laser pumps a non-linear crystal. With some probability a photon in the pump beam will be split into two photons with orthogonal polarisation $|\updownarrow\rangle$ and $|\leftrightarrow\rangle$ along the surface of the two respective cones. Depending on the optical axis of the crystal, the two cones are slightly tilted from each other. Selecting the spatial modes at the intersection of the two cones yields the outgoing state $|0\rangle + \xi|\Psi^-\rangle + O(\xi^2)$.

the direction of the pump). In the two spatial modes where the different polarisation cones intersect we can no longer infer the polarisation of the photons, and as a consequence the two photons become entangled in their polarisation.

However, parametric down-converters do not produce Bell states [8,11,12]. They form a class of devices yielding Gaussian evolutions:

$$|\Psi\rangle = U(t)|0\rangle = \exp[-it\mathcal{H}_I/\hbar]|0\rangle\ , \tag{6}$$

with

$$\mathcal{H}_I = \frac{1}{2}\sum_{ij} \hat{a}_i^\dagger \Lambda_{ij} \hat{a}_j^\dagger + \text{H.c.}\ , \tag{7}$$

where $\mathcal{H}_I$ is the interaction Hamiltonian, $\hat{a}_i^\dagger$ a creation operator and Λ_{ij} the components of a (complex) symmetric matrix. Here, H.c. stands for the Hermitian conjugate. If Λ is diagonal the evolution U corresponds to a collection of single-mode squeezers [13]. In the case of degenerate type II parametric down-conversion used to produce a photon-pair exhibiting polarisation entanglement, the interaction Hamiltonian in the rotating-wave approximation is

$$\mathcal{H}_I = \kappa(\hat{a}_x^\dagger \hat{b}_y^\dagger - \hat{a}_y^\dagger \hat{b}_x^\dagger) + \text{H.c.}\ , \tag{8}$$

with κ a parameter which is determined by the strength of the pump and the coupling of the electro-magnetic field to the crystal.

The outgoing state of the down-converter is then

$$|\Psi\rangle_{ab} = \left(1-\xi^2\right)|0,0\rangle_{ab} + \xi\Big(|x,y\rangle_{ab} - |y,x\rangle_{ab}\Big)$$

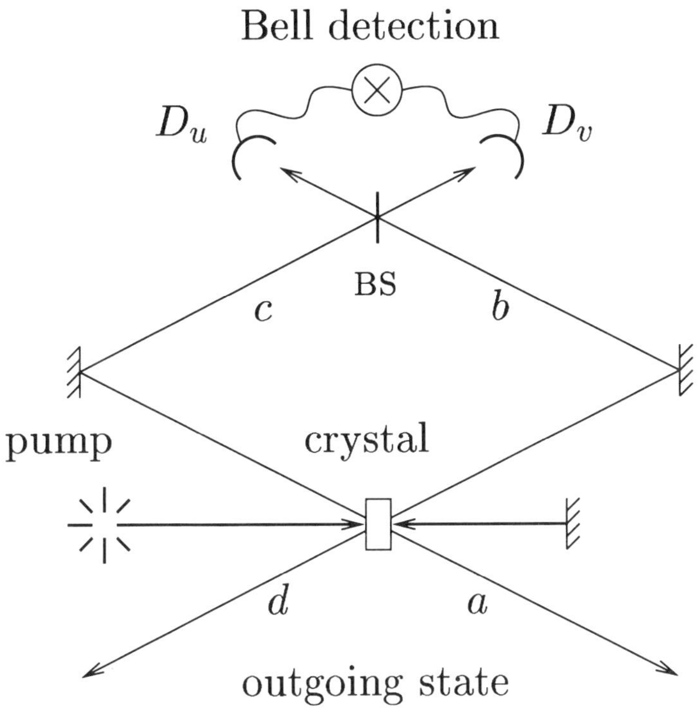

Fig. 2. A schematic representation of the experimental setup for entanglement swapping as performed by Pan *et al.* The pump beam is reverted by a mirror in order to create two entangled photon-pairs in different directions (modes a, b, c and d). Modes b and c are sent into a beam-splitter (BS). A coincidence in the detectors D_u and D_v at the outgoing modes of the beam-splitter identify a $|\Psi^-\rangle$ Bell state. The undetected modes a and d are now in the $|\Psi^-\rangle$ Bell state as well.

$$+\xi^2\Big(|x^2,y^2\rangle_{ab} - |xy,xy\rangle_{ab} + |y^2,x^2\rangle_{ab}\Big) + O(\xi^3)\ , \qquad (9)$$

with $\xi \ll 1$, which is a function of κ. Here, $|x^2\rangle$ denotes an x-polarised mode in a 2-photon Fock state (the case of two y-polarised or an x- and a y-polarised photon are treated similarly).

For the experimental demonstration of entanglement swapping we need two independent Bell states. A Bell measurement on one half of either Bell state will then entangle the two remaining modes. The photons in these modes do not originate from a common source, i.e., they have never interacted. Yet they are now entangled.

In the experiment conducted by Pan *et al.*, instead of having two parametric down-converters, one crystal was pumped twice in opposite directions (see Fig. 2). This way, a state which is equivalent to a state originating from two independent down-converters was obtained. In order to simplify our dis-

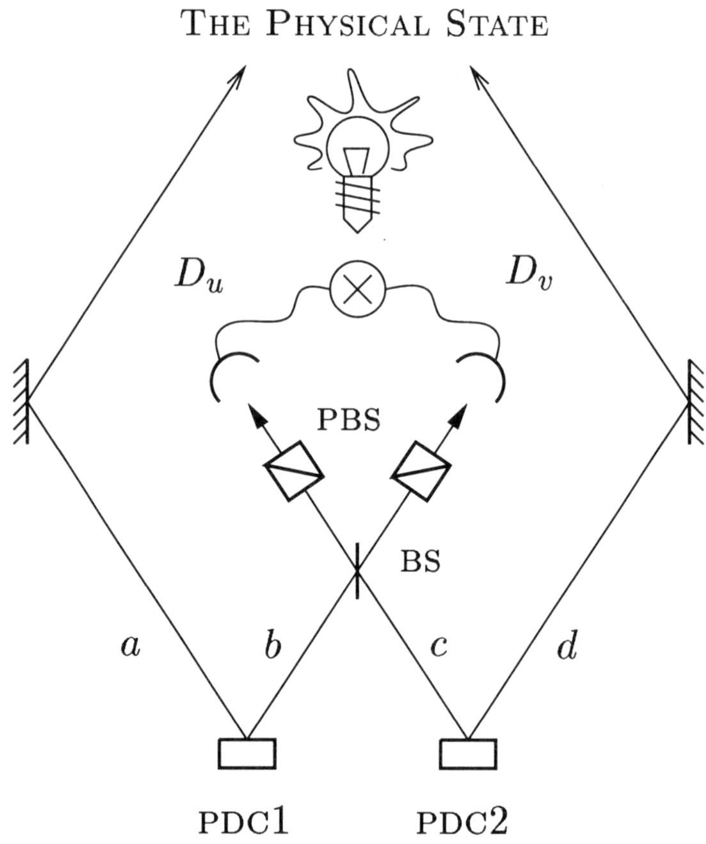

Fig. 3. A schematic representation of the entanglement swapping setup. Two parametric down-converters (PDC) create states which exhibit polarisation entanglement. One branch of each source is sent into a beam splitter (BS), after which the polarisation beam splitters (PBS) select particular polarisation settings. A coincidence in detectors D_u and D_v ideally identify the $|\Psi^-\rangle$ Bell state. However, since there is a possibility that one down-converter produces two photon-pairs while the other produces nothing, the detectors D_u and D_v no longer constitute a Bell-detection, and the freely propagating PHYSICAL STATE is no longer a pure Bell state.

cussion, we will treat the experimental setup as if it consists of two separate down-converters (see Fig. 3).

One spatial mode of either down-conversion state is sent into a beam-splitter, the output of which is detected. This constitutes the Bell measurement. In the case where both down-converters create a polarisation entangled photon-pair, a coincidence in the photo-detectors D_u and D_v identify the antisymmetric Bell state $|\Psi^-\rangle$ [14]. The outgoing state should then be the $|\Psi^-\rangle$ Bell state as well, thus creating event-ready entanglement.

However, there are two problems. First, this is not a complete Bell measurement [15,16], i.e., it is not possible to identify all the four Bell states simultaneously. The consequence is that entanglement swapping occurs a quarter of the time (only $|\Psi^-\rangle$ is identified).

A second, and more serious, problem is that when we study coincidences between two down-converters, we need to take higher-order photon-pair production into account [see Eq. (9)]. For instance, one down-converter creates a photon-pair with probability $|\xi|^2$. Two down-converters therefore create two photon-pairs with probability $|\xi|^4$. However, this is roughly equal to the probability where one down-converter produces nothing (i.e., the vacuum $|0\rangle$), while the other produces two photon-pairs. In the next section we will show that for this reason a coincidence in the detectors D_u and D_v no longer uniquely identifies a $|\Psi^-\rangle$ Bell state.

3 Event-Ready Entanglement

In this section we first investigate the effect of double-pair production on the Bell measurement in the experimental setup depicted in Fig. 3. Subsequently, we study the possibility of event-ready entanglement in the context of entanglement swapping.

There is a possibility that a single down-converter produces a double photon-pair. This means that the two photons in the outgoing modes of the beam-splitter in Fig. 3 do not necessarily originate from different down-converters. We can therefore no longer interpret a detector-coincidence at the outgoing modes of the beam-splitter as a projection onto the $|\Psi^-\rangle$ Bell state.

Another way of looking at this is as follows: consider a two-photon polarisation state. It is a vector in a Hilbert space generated by (for instance) the basis vectors $|x,x\rangle$, $|x,y\rangle$, $|y,x\rangle$ and $|y,y\rangle$. The $|\Psi^-\rangle$ Bell state is a superposition of these basis vectors. The key observation is that the two photons described in this Hilbert space occupy *different spatial modes*. When a two-photon state entering a 50:50 beam-splitter gives a two-fold coincidence, this state is projected onto the $|\Psi^-\rangle$ Bell state.

This Hilbert space should be clearly distinguished from a (truncated) Fock space. In the Fock space two photons can occupy the same spatial mode (see for example the state in Eq. (9)). As a consequence, the two input modes of a beam-splitter can be the vacuum $|0\rangle$ and a two-photon state (for instance $|x^2\rangle$) respectively. In this scenario a detector coincidence at the output of the beam-splitter is still possible, but it can *not* be interpreted as the projection of the incoming state $|0,x^2\rangle$ on the $|\Psi^-\rangle$ Bell state (see Fig. 4).

In the case of the entanglement swapping experiment, two photon-pairs are created either by one down-converter alone or both down-converters. This means that, conditioned on a detector coincidence, the state entering the beam-splitter is a superposition of two single-photon states plus the vacuum

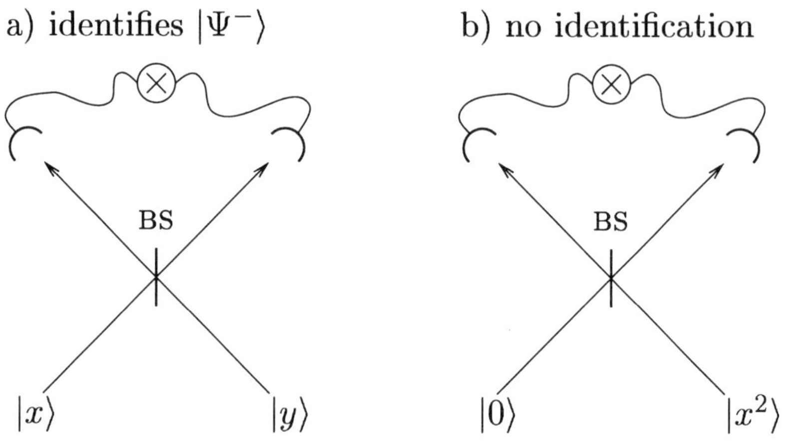

Fig. 4. A schematic representation of the Bell measurement in the entanglement swapping experiment. In Fig. 4a both incoming modes are occupied by a single photon. A coincidence in the detectors will then identify a projection onto the $|\Psi^-\rangle$ Bell state. In Fig. 4b, however, one input mode is the vacuum, whereas the other is populated by two photons. In this case a detector coincidence cannot be interpreted as a projection onto the $|\Psi^-\rangle$ Bell state. Both instances a) and b) occur in the entanglement swapping experiment.

and a two-photon state. In this setup a detector coincidence therefore does not identify the $|\Psi^-\rangle$ Bell state.

In Fig. 3 we have added two polarisation beam-splitters in the outgoing modes of the beam-splitter. This allows us to condition the outgoing state in modes a and d on the different polarisation settings. It should be noted that for a Bell detection depicted in Fig. 4a we do not need a polarisation sensitive measurement. However, since the detector coincidence is no longer a Bell measurement it will be convenient to distinguish between the four different polarisation settings at the output modes of the beam-splitter [(x,x), (x,y), (y,x) and (y,y)].

It turns out that the four outgoing states conditioned on the four different polarisation settings have a remarkably simple form:

$$\begin{aligned}
|\phi_{(x,x)}\rangle_{\rm ad} &= \big(|0,y^2\rangle - |y^2,0\rangle\big)/\sqrt{2}, \\
|\phi_{(x,y)}\rangle_{\rm ad} &= \big(|0,xy\rangle - |y,x\rangle + |x,y\rangle - |xy,0\rangle\big)/2, \\
|\phi_{(y,x)}\rangle_{\rm ad} &= \big(|0,xy\rangle + |y,x\rangle - |x,y\rangle - |xy,0\rangle\big)/2, \\
|\phi_{(y,y)}\rangle_{\rm ad} &= \big(|0,x^2\rangle - |x^2,0\rangle\big)/\sqrt{2}\,. \qquad (10)
\end{aligned}$$

These states can also be obtained by sending $|y,y\rangle$, $|y,x\rangle$, $|x,y\rangle$ and $|x,x\rangle$ into a 50:50 beam-splitter respectively (see Fig. 5). When no distinction between

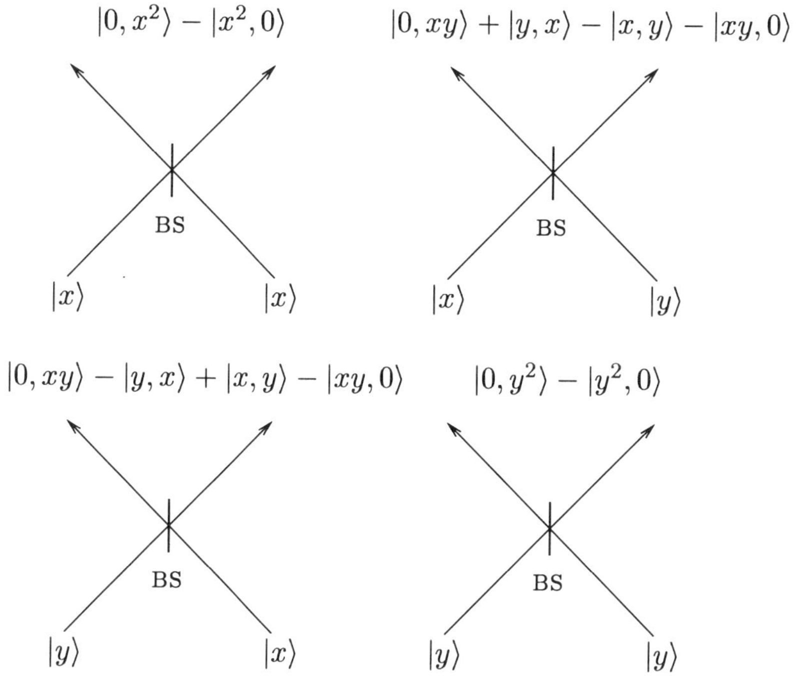

Fig. 5. The four (unnormalised) outgoing states from a 50:50 beam-splitter conditioned on the four input states $|x,x\rangle$, $|x,y\rangle$, $|y,x\rangle$ and $|y,y\rangle$. The outgoing states correspond to the four possible outgoing states of entanglement swapping.

the four possible polarisation settings in the two-fold detector-coincidence is made [10,8], the state of the two remaining (undetected) modes (the physical state) will be a mixture ρ of the four states in Eqs. (10).

Using a technical mathematical criterion based on the partial transpose of the outgoing state ρ [17,18], it can be shown that ρ is entangled (ρ satisfies this entanglement criterion). However, it can not be used for event-ready detections of polarisation entanglement since the states in Eqs. (10) are not of the form of Eq. (3). More specifically, these states are nowhere near the maximally entangled states we wanted to create.

We now ask the question whether entanglement swapping can still be used to create event-ready entanglement. Observe that the experiment conducted by Pan *et al.* closely resembles the experimental realisation of quantum teleportation by Bouwmeester *et al.* [4]. There, one of the states produced by parametric down-conversion was used to prepare a single-photon state which was then teleported. However, due to the double pair-production, the teleported state was a mixture of the teleported photon and the vacuum.

In Ref. [11], Braunstein and Kimble presented three possible ways to improve the experimental setup of quantum teleportation in order to minimise the unwanted double pairs created in a single down-converter. Since the experimental setup considered here closely resembles the setup of the teleportation experiment, one might expect that the remedy given by Braunstein and Kimble will be effective here as well.

However, this is not the case. The first approach was to make sure that in one of the outgoing modes (the state-preparation mode) there was only one photon present. In order to achieve this, a detector cascade was suggested which, upon a two-fold detector coincidence, would reveal a two-photon state. Since in the entanglement swapping experiment there is no *conditional* measurement of the outgoing modes (apart from the Bell measurement), this approach doesn't work here.

The second approach is to differentiate between the coupling of the two down-converters by lowering the intensity of the pump of one of them. However, since the setup of the entanglement swapping experiment is symmetric, it can be easily seen that the (unnormalised) outgoing states become

$$\begin{aligned}
|\phi_{(x,x)}\rangle &\propto \xi_2^2\,|0,y^2\rangle - \xi_1^2\,|y^2,0\rangle + O(\xi^3), \\
|\phi_{(x,y)}\rangle &\propto \xi_2^2\,|0,xy\rangle - \xi_1\xi_2\,(|y,x\rangle - |x,y\rangle) - \xi_1^2\,|xy,0\rangle + O(\xi^3), \\
|\phi_{(y,x)}\rangle &\propto \xi_2^2\,|0,xy\rangle + \xi_1\xi_2\,(|y,x\rangle - |x,y\rangle) - \xi_1^2\,|xy,0\rangle + O(\xi^3), \\
|\phi_{(y,y)}\rangle &\propto \xi_2^2\,|0,x^2\rangle - \xi_1^2\,|x^2,0\rangle + O(\xi^3)\,.
\end{aligned} \tag{11}$$

Varying ξ_1 and ξ_2 will *not* allow us to create any of the states which have the form of Eq. (3).

The third approach presented by Braunstein and Kimble involved the use of quantum non-demolition (QND) measurements. In the case of entanglement swapping, we should be able to identify whether there are photons in the outgoing modes with such QND detectors. If we can tell that both outgoing modes are populated by a photon without destroying the non-local correlations (i.e., without gaining information about the polarisations of the photons), we can create event-ready entanglement. However, such QND detectors correspond to technology not yet available for optical photons (for QND detections of microwave photons see Ref. [19]).

Can we turn any of the states in Eq. (10) into the form of Eq. (3) by other means? Additional photon sources would take us beyond the entanglement swapping protocol and we will not consider them here. Alternatively, we might be able to use a linear interferometer to obtain event-ready entanglement. First, observe that instead of taking Eqs. (10) as the input of the interferometer, we can use $|x,x\rangle$, $|x,y\rangle$, $|y,x\rangle$ or $|y,y\rangle$, since these states are obtained from Eqs. (10) by means of a (unitary) beam-splitter operation. If the linear interferometer described above does not exist for these separable states, neither will it for the outgoing states of the entanglement swapping protocol, since these states only differ by a unitary transformation.

Next, suppose we have a linear interferometer described by the unitary matrix U [20], which transforms the creation operators of the electro-magnetic field according to

$$\hat{a}_i \to \sum_j u_{ij}\hat{b}_j,$$
$$\hat{a}_i^\dagger \to \sum_j u_{ij}^*\hat{b}_j^\dagger\,, \tag{12}$$

where the u_{ij} are the components of U and i, j enumerate both the modes and polarisations. There is no mixing between the creation and annihilation operators, because photons do not interact with each other.

In order to create event-ready entanglement, the creation operators which give rise to the states obtained by the entanglement swapping protocol (for instance $|x, y\rangle = a_x^\dagger b_y^\dagger |0\rangle$) should be transformed according to

$$\hat{a}_x^\dagger \hat{b}_y^\dagger \;\to\; (\hat{c}_x^\dagger \hat{d}_y^\dagger - \hat{c}_y^\dagger \hat{d}_x^\dagger)/\sqrt{2}\,. \tag{13}$$

Relabel the modes a_x, a_y, b_x and b_y as a_1 to a_4, respectively. Without loss of generality (and leaving the normalisation aside for the moment) we can then write

$$\hat{a}_1^\dagger \hat{a}_2^\dagger \;\to\; \hat{b}_1^\dagger \hat{b}_2^\dagger - \hat{b}_3^\dagger \hat{b}_4^\dagger\,. \tag{14}$$

Substituting Eq. (12) into Eq. (14) generates ten equations for eight variables u_{ij}^*:

$$\begin{aligned}
u_{11}^* u_{21}^* = u_{12}^* u_{22}^* = u_{13}^* u_{23}^* = u_{14}^* u_{24}^* \quad &= 0,\\[1ex]
(u_{11}^* u_{23}^* + u_{13}^* u_{21}^*) = (u_{11}^* u_{24}^* + u_{14}^* u_{21}^*) &= 0,\\
(u_{12}^* u_{23}^* + u_{13}^* u_{22}^*) = (u_{12}^* u_{24}^* + u_{14}^* u_{22}^*) &= 0,\\[1ex]
(u_{11}^* u_{22}^* + u_{12}^* u_{21}^*) \quad &= 1,\\
(u_{13}^* u_{24}^* + u_{14}^* u_{23}^*) \quad &= -1\,.
\end{aligned} \tag{15}$$

It can be easily verified that there are no solutions for the u_{ij}^* which satisfy these ten equations simultaneously. This means that there is no passive interferometer which transforms the states resulting from entanglement swapping into event-ready entangled states.

4 Conclusions

We speak of event-ready entanglement when upon the production of a maximally polarisation entangled state the source flashes a red light (or gives a similar macroscopic indication). We have studied the possibility of event-ready entanglement preparation by means of entanglement swapping as performed by Pan *et al.* [10]. This was suggested by Zukowski, Zeilinger, Horne and Ekert [8] and Pavičić [9].

In entanglement swapping, we perform a Bell measurement on two parts of two (maximally) entangled states, leaving the two undetected parts entangled. In general, these parts have never interacted. In the experiment of Pan *et al.*, the entangled states are produced by means of parametric down-conversion. Due to higher order corrections in the down-converters, the physical state leaving the entanglement swapping apparatus is a random mixture of four states. These states correspond to the four possible polarisation settings in the Bell measurement. They are equivalent to the outgoing state of a 50:50 beam-splitter conditioned on the four possible input states $|x,x\rangle$, $|x,y\rangle$, $|y,x\rangle$ and $|y,y\rangle$. However, these states are not the ones we were looking for.

With respect to a related experiment involving quantum teleportation [4], it has been described [11] how to enhance the fidelity of the outgoing state. However, these methods fail here, as well as the application of a linear interferometer with passive elements. We need at least a quantum non-demolition measurement or a quantum computer of some kind to turn the outgoing states of the entanglement swapping experiment into event-ready entanglement.

Appendix: Transformation of Maximally Entangled States

In this appendix we will show that each maximally entangled two-system state can be transformed into any other by a local unitary transformation on one subsystem alone. For example, every two-photon polarisation Bell state can be transformed into *any* other maximally polarisation entangled two-photon state by a linear optical transformation on one mode.

We will treat this in a formal way by considering an arbitrary maximally entangled state of two N-level systems in the Schmidt decomposition:

$$|\psi\rangle = \frac{1}{\sqrt{N}} \sum_{j=1}^{N} e^{i\phi_j} |n_j, m_j\rangle \; , \tag{16}$$

i.e., a state with equal amplitudes on all possible branches. Here $\{|n_i\rangle\}$ and $\{|m_i\rangle\}$ form two orthonormal bases of the subsystems. We can obtain *any* maximally entangled state by applying the (bi-local) unitary transformation $U_1 \otimes U_2$. This means that each maximally entangled state can be transformed into any other by a pair of local unitary transformations on each of the subsystems. We will now show that any two maximally entangled states $|\psi\rangle$ and $|\psi'\rangle$ are connected by a local unitary transformation on one subsystem alone:

$$|\psi\rangle = U \otimes \mathbb{1} \, |\psi'\rangle \; . \tag{17}$$

First, we show that any transformation $U_1 \otimes U_2$ on a *particular* maximally entangled state $|\phi\rangle$ can be written as $V \otimes \mathbb{1}|\phi\rangle$, where $V = U_1 U_2^T$. Take

$$|\phi\rangle = \frac{1}{\sqrt{N}} \sum_{j=1}^{N} |n_j, m_j\rangle \; . \tag{18}$$

For any U we have

$$U \otimes U^* |\phi\rangle = |\phi\rangle \ . \tag{19}$$

Proof. Using the completeness relation

$$\sum_{k=1}^{N} |n_k\rangle\langle n_k| = \mathbb{1} \tag{20}$$

on both subsystems we have

$$U \otimes U^* |\phi\rangle = \sum_{k,l=1}^{N} |n_k, m_l\rangle\langle n_k, m_l| \left(U \otimes \mathbb{1}\right)\left(\mathbb{1} \otimes U^*\right)|\phi\rangle \ . \tag{21}$$

By writing out $|\phi\rangle$ explicitly according to Eq. (18) we obtain

$$\begin{aligned} U \otimes U^* |\phi\rangle &= \frac{1}{\sqrt{N}} \sum_{j,k,l=1}^{N} U_{kj} U^*_{lj} |n_k, m_l\rangle \\ &= \frac{1}{\sqrt{N}} \sum_{k,l=1}^{N} (UU^\dagger)_{kl} |n_k, m_l\rangle \\ &= \frac{1}{\sqrt{N}} \sum_{k,l=1}^{N} \delta_{kl} |n_k, m_l\rangle \ , \end{aligned} \tag{22}$$

which is equal to $|\phi\rangle$. □

Next, using the relation (19) we will show that every transformation $U_1 \otimes U_2$ acting on $|\phi\rangle$ is equivalent to a transformation $V \otimes \mathbb{1}$ acting on $|\phi\rangle$, where $V = U_1 U_2^T$.

Proof. From Eq. (19) we obtain the equation

$$\left(\mathbb{1} \otimes U^T\right)\left(U \otimes U^*\right)|\phi\rangle = \left(U \otimes \mathbb{1}\right)|\phi\rangle \ , \tag{23}$$

which immediately gives us

$$\begin{aligned} U \otimes \mathbb{1} |\phi\rangle &= \mathbb{1} \otimes U^T |\phi\rangle \qquad \text{and} \\ U^T \otimes \mathbb{1} |\phi\rangle &= \mathbb{1} \otimes U |\phi\rangle \ . \end{aligned} \tag{24}$$

From Eq. (24) we obtain

$$\begin{aligned} U_1 \otimes U_2 |\phi\rangle &= \left(U_1 \otimes \mathbb{1}\right)\left(\mathbb{1} \otimes U_2\right)|\phi\rangle \\ &= \left(U_1 \otimes \mathbb{1}\right)\left(U_2^T \otimes \mathbb{1}\right)|\phi\rangle \\ &= U_1 U_2^T \otimes \mathbb{1} |\phi\rangle \ . \end{aligned} \tag{25}$$

Similarly,

$$U_1 \otimes U_2|\phi\rangle = \mathbb{1} \otimes U_1^T U_2|\phi\rangle \; . \tag{26}$$

We therefore obtain that $U_1 \otimes U_2|\phi\rangle$ is equal to $V \otimes \mathbb{1}|\phi\rangle$ with $V = U_1 U_2^T$, and similarly that it is equal to $\mathbb{1} \otimes V'|\phi\rangle$ with $V' = U_1^T U_2$. □

Since every maximally entangled stated can be obtained by applying $U_1 \otimes U_2$ to $|\phi\rangle$, two maximally entangled states $|\psi\rangle$ and $|\psi'\rangle$ can be transformed into any other by choosing

$$\begin{aligned} |\psi\rangle &= U \otimes \mathbb{1}\, |\phi\rangle \\ |\psi'\rangle &= V \otimes \mathbb{1}\, |\phi\rangle \;\; , \end{aligned} \tag{27}$$

which gives

$$|\psi\rangle = UV^\dagger \otimes \mathbb{1}\, |\psi'\rangle \; . \tag{28}$$

Thus each maximally entangled two-system state can be obtained from any other by means of a local unitary transformation on one subsystem alone.

Returning to a pair of maximally polarisation entangled photons, we know that any unitary transformation on a single optical mode consisting of single photons can be viewed as a combination of a polarisation rotation and a polarisation dependent phase shift of that mode. We can easily perform such operations with linear optical elements. If we can create one maximally entangled state, we can create *any* maximally entangled state. It is therefore sufficient to restrict our discussion to, for example, the $|\Psi^-\rangle$ Bell state.

References

1. Einstein A., Podolsky B. and Rosen N.(1935): Phys. Rev. **47**, 777.
2. Bell J. S. (1964): Phys. **1**, 195; also in (1987) *Speakable and unspeakable in quantum mechanics*, Cambridge University Press, Cambridge.
3. Bennett C. H., Brassard G., Crépeau C., Jozsa R., Peres A., Wootters W. K. (1993): Phys. Rev. Lett. **70**, 1895.
4. Bouwmeester D., Pan J.-W., Mattle K., Eibl M., Weinfurter H., Zeilinger A. (1997): Nature **390**, 575.
5. Boschi D., Branca S., De Martini F., Hardy L., and Popescu S. (1998): Phys. Rev. Lett. **80**, 1121.
6. Furusawa A., Sørensen J. L., Braunstein S. L., Fuchs C. A., Kimble H. J., Polzik E. S. (1998): Science **282**, 706.
7. Kwiat P. G., Mattle K., Weinfurter H., Zeilinger A. (1995): Phys. Rev. Lett. **75**, 4337.
8. Zukowski M., Zeilinger A., Horne M. A., Ekert A. K. (1993): Phys. Rev. Lett. **71**, 4287.
9. Pavičić M. (1996): Event-ready entanglement preparation. In: De Martini F., Denardo G., Shih Y. (Eds.) *Quantum Interferometry*, VCH Publishing Division I, New York.

10. Pan J.-W., Bouwmeester D., Weinfurter H., Zeilinger A. (1998): Phys. Rev. Lett. **80**, 3891.
11. Braunstein S. L., Kimble H. J. (1998): Nature **394**, 840.
12. Kok P., Braunstein S. L. (1999): *Post-Selected versus Non-Post-Selected Quantum Teleportation using Parametric Down-Conversion.* LANL e-print quant-ph/9903074.
13. Braunstein S. L. (1999): *Squeezing as an irreducible resource.* LANL e-print quant-ph/9904002.
14. Braunstein S. L., Mann A. (1995): Phys. Rev. A, **51**, R1727.
15. Lütkenhaus N., Calsamiglia J., Suominen K.-A. (1999): Phys. Rev. A. **59**, 3295.
16. Vaidman L., Yoran N. (1999): Phys. Rev. A. **59**, 116.
17. Peres A. (1996): Phys. Rev. Lett. **77**, 1413.
18. Horodecki M., Horodecki P., Horodecki R. (1996): Phys. Lett. A. **223**, 1.
19. Nogues G., Rauschenbeutel A., Osnaghi S., Brune M., Raimond J. M., Haroche S. (1999): Nature **400**, 239.
20. Reck M., Zeilinger A., Bernstein H. J., Bertani P. (1994): Phys. Rev. Lett. **73**, 58.

Radiation Damping and Decoherence in Quantum Electrodynamics

Heinz–Peter Breuer[1] and Francesco Petruccione[1,2]

[1] Fakultät für Physik, Albert-Ludwigs-Universität Freiburg, Hermann–Herder–Str. 3, D–79104 Freiburg i. Br., Germany
[2] Istituto Italiano per gli Studi Filosofici, Palazzo Serra di Cassano, Via Monte di Dio 14, I–80132 Napoli, Italy

Abstract. The processes of radiation damping and decoherence in Quantum Electrodynamics are studied from an open system's point of view. Employing functional techniques of field theory, the degrees of freedom of the radiation field are eliminated to obtain the influence phase functional which describes the reduced dynamics of the matter variables. The general theory is applied to the dynamics of a single electron in the radiation field. From a study of the wave packet dynamics a quantitative measure for the degree of decoherence, the decoherence function, is deduced. The latter is shown to describe the emergence of decoherence through the emission of bremsstrahlung caused by the relative motion of interfering wave packets. It is argued that this mechanism is the most fundamental process in Quantum Electrodynamics leading to the destruction of coherence, since it dominates for short times and because it is at work even in the electromagnetic field vacuum at zero temperature. It turns out that decoherence trough bremsstrahlung is very small for single electrons but extremely large for superpositions of many-particle states.

1 Introduction

Decoherence may be defined as the (partial) destruction of quantum coherence through the interaction of a quantum mechanical system with its surroundings. In the theoretical analysis decoherence can be studied with the help of simple microscopic models which describe, for example, the interaction of a quantum mechanical system with a collection of an infinite number of harmonic oscillators, representing the environmental degrees of freedom [1,2]. In an open system's approach to decoherence one derives dynamic equations for the reduced density matrix [3] which yields the state of the system of interest as it is obtained from an average over the degrees of freedom of the environment and the resulting loss of information on the entangled state of the combined total system. The strong suppression of coherence can then be explained by showing that the reduced density matrix equation leads to an extremely rapid transitions of a coherent superposition to an incoherent statistical mixture [4,5]. For certain superpositions the associated decoherence time scale is often found to be smaller than the corresponding relaxation or damping time by many orders of magnitude. This is a signature for the fundamental distinction between the notions of decoherence and of dissipation.

A series of interesting experimental investigations of decoherence have been performed as, for example, experiments on Schrödinger cat states of a cavity field mode [6] and on single trapped ions in a controllable environment [7].

If one considers the coherence of charged matter, it is the electromagnetic field which plays the rôle of the environment. It is the purpose of this paper to study the emergence of decoherence processes in Quantum Electrodynamics (QED) from an open system's point of view, that is by an elimination of the degrees of freedom of the radiation field. An appropriate technique to achieve this goal is the use of functional methods from field theory. In section 2 we combine these methods with a super-operator approach to derive an exact, relativistic representation for the reduced density matrix of the matter degrees of freedom. This representation involves an influence phase functional that completely describes the influence of the electromagnetic radiation field on the matter dynamics. The influence phase functional may be viewed as a super-operator representation of the Feynman-Vernon influence phase [1] which is usually obtained with the help of path integral techniques.

In section 3 we treat the problem of a single electron in the radiation field within the non-relativistic approximation. Starting from the influence phase functional, we formulate the reduced electron motion in terms of a path integral which involves an effective action functional. The corresponding classical equations of motion are demonstrated to yield the Abraham-Lorentz equation describing the radiation damping of the electron motion. In addition, the influence phase is shown to lead to a decoherence function which provides a measure for the degree of decoherence.

The general theory will be illustrated with the help of two examples, namely a free electron (section 4) and an electron moving in a harmonic potential (section 5). For both cases an analytical expression for the decoherence function is found, which describes how the radiation field affects the electron coherence.

We shall use the obtained expressions to investigate in detail the time-evolution of Gaussian wave packets. We study the influence of the radiation field on the interference pattern which results from the collision of two moving wave packets of a coherent superposition. It turns out that the basic mechanism leading to the decoherence of matter waves is the emission of bremsstrahlung through the moving wave packets. The resultant picture of decoherence is shown to yield expressions for the decoherence time and length scales which differ substantially from the conventional estimates derived from the prominent Caldeira-Leggett master equation. In particular, it will be shown that a superposition of two wave packets with zero velocity does not decohere and, thus, the usual picture of decoherence as a decay of the off-diagonal peaks in the corresponding density matrix does not apply to decoherence through bremsstrahlung.

We investigate in section 6 the possibility of the destruction of coherence of the superposition of many-particle states. It will be argued that, while the decoherence effect is small for single electrons at non-relativistic speed, it is drastically amplified for certain superpositions of many-particle states.

Finally, we draw our conclusions in section 7.

2 Reduced Density Matrix of the Matter Degrees of Freedom

Our aim is to eliminate the variables of the electromagnetic radiation field to obtain an exact representation for the reduced density matrix ρ_m of the matter degrees of freedom. The starting point will be the following formal equation which relates the density matrix $\rho_m(t_f)$ of the matter at some final time t_f to the density matrix $\rho(t_i)$ of the combined matter-field system at some initial time t_i,

$$\rho_m(t_f) = \mathrm{tr}_f \left\{ T_{\leftarrow} \exp\left[\int_{t_i}^{t_f} \mathrm{d}^4 x \mathcal{L}(x) \right] \rho(t_i) \right\}. \tag{1}$$

The Liouville super-operator $\mathcal{L}(x)$ is defined as

$$\mathcal{L}(x)\rho \equiv -i[\mathcal{H}(x), \rho], \tag{2}$$

where $\mathcal{H}(x)$ denotes the Hamiltonian density. Space-time coordinates are written as $x^\mu = (x^0, \boldsymbol{x}) = (t, \boldsymbol{x})$, where the speed of light c is set equal to 1. All fields are taken to be in the interaction picture and $T_{\leftarrow}$ indicates the chronological time-ordering of the interaction picture fields, while tr_f denotes the trace over the variables of the radiation field. Setting $\hbar = c = 1$ we shall use here Heaviside-Lorentz units such that the fine structure constant is given by

$$\alpha = \frac{e^2}{4\pi\hbar c} \approx \frac{1}{137}. \tag{3}$$

To be specific we choose the Coulomb gauge in the following which means that the Hamiltonian density takes the form [8–10]

$$\mathcal{H}(x) = \mathcal{H}_{\mathrm{C}}(x) + \mathcal{H}_{\mathrm{tr}}(x). \tag{4}$$

Here,

$$\mathcal{H}_{\mathrm{tr}}(x) = j^\mu(x) A_\mu(x) \tag{5}$$

represents the density of the interaction of the matter current density $j^\mu(x)$ with the transversal radiation field,

$$A^\mu(x) = (0, \boldsymbol{A}(x)), \qquad \boldsymbol{\nabla} \cdot \boldsymbol{A}(x) = 0, \tag{6}$$

and

$$\mathcal{H}_{\mathrm{C}}(x) = \frac{1}{2} j^0(x) A^0(x) = \frac{1}{2} \int \mathrm{d}^3 y \, \frac{j^0(x^0, \boldsymbol{x}) j^0(x^0, \boldsymbol{y})}{4\pi |\boldsymbol{x} - \boldsymbol{y}|} \tag{7}$$

is the Coulomb energy density such that

$$H_C(x^0) = \frac{1}{2}\int \mathrm{d}^3x \int \mathrm{d}^3y \frac{j^0(x^0,\boldsymbol{x})j^0(x^0,\boldsymbol{y})}{4\pi|\boldsymbol{x}-\boldsymbol{y}|} \tag{8}$$

is the instantaneous Coulomb energy. Note that we use here the convention that the electron charge e is included in the current density $j^\mu(x)$ of the matter.

Our first step is a decomposition of chronological time-ordering operator $T_\leftarrow$ into a time-ordering operator $T^j_\leftarrow$ for the matter current and a time-ordering operator $T^A_\leftarrow$ for the electromagnetic field,

$$T_\leftarrow = T^j_\leftarrow T^A_\leftarrow. \tag{9}$$

This enables one to write Eq. (1) as

$$\rho_m(t_f) = T^j_\leftarrow \left(\mathrm{tr}_f \left\{ T^A_\leftarrow \exp\left[\int_{t_i}^{t_f} \mathrm{d}^4x \left(\mathcal{L}_\mathrm{C}(x) + \mathcal{L}_\mathrm{tr}(x)\right)\right] \rho(t_i) \right\} \right), \tag{10}$$

where we have introduced the Liouville super-operators for the densities of the Coulomb field and of the transversal field,

$$\mathcal{L}_\mathrm{C}(x)\rho \equiv -\mathrm{i}[\mathcal{H}_\mathrm{C}(x),\rho], \qquad \mathcal{L}_\mathrm{tr}(x)\rho \equiv -\mathrm{i}[j^\mu(x)A_\mu(x),\rho]. \tag{11}$$

The currents j^μ commute under the time-ordering $T^j_\leftarrow$. We may therefore treat them formally as commuting c-number fields under the time-ordering symbol. Since the super-operator $\mathcal{L}_\mathrm{C}(x)$ only contains matter variables, the corresponding contribution can be pulled out of the trace. Hence, we have

$$\rho_m(t_f) = T^j_\leftarrow \left(\exp\left[\int_{t_i}^{t_f} \mathrm{d}^4x \mathcal{L}_\mathrm{C}(x)\right] \mathrm{tr}_f \left\{ T^A_\leftarrow \exp\left[\int_{t_i}^{t_f} \mathrm{d}^4x \mathcal{L}_\mathrm{tr}(x)\right] \rho(t_i) \right\} \right). \tag{12}$$

We now proceed by eliminating the time-ordering of the A-fields. With the help of the Wick-theorem we get

$$T^A_\leftarrow \exp\left[\int_{t_i}^{t_f} \mathrm{d}^4x \mathcal{L}_\mathrm{tr}(x)\right] = \tag{13}$$
$$\exp\left[\frac{1}{2}\int_{t_i}^{t_f} \mathrm{d}^4x \int_{t_i}^{t_f} \mathrm{d}^4x' [\mathcal{L}_\mathrm{tr}(x),\mathcal{L}_\mathrm{tr}(x')]\theta(t-t')\right] \exp\left[\int_{t_i}^{t_f} \mathrm{d}^4x \mathcal{L}_\mathrm{tr}(x)\right].$$

In order to determine the commutator of the Liouville super-operators we invoke the Jacobi identity which yields for an arbitrary test density ρ,

$$\begin{aligned} [\mathcal{L}_\mathrm{tr}(x),\mathcal{L}_\mathrm{tr}(x')]\rho &= \mathcal{L}_\mathrm{tr}(x)\mathcal{L}_\mathrm{tr}(x')\rho - \mathcal{L}_\mathrm{tr}(x')\mathcal{L}_\mathrm{tr}(x)\rho \\ &= -[\mathcal{H}_\mathrm{tr}(x),[\mathcal{H}_\mathrm{tr}(x'),\rho]] + [\mathcal{H}_\mathrm{tr}(x'),[\mathcal{H}_\mathrm{tr}(x),\rho]] \\ &= -[[\mathcal{H}_\mathrm{tr}(x),\mathcal{H}_\mathrm{tr}(x')],\rho]. \end{aligned} \tag{14}$$

The commutator of the transversal energy densities may be simplified to read

$$[\mathcal{H}_{\rm tr}(x), \mathcal{H}_{\rm tr}(x')] = j^{\mu}(x) j^{\nu}(x') [A_{\mu}(x), A_{\nu}(x')], \tag{15}$$

since the contribution involving the commutator of the currents vanishes by virtue of the time-ordering operator $T^j_{\leftarrow}$. Thus, it follows from Eqs. (14) and (15) that the commutator of the Liouville super-operators may be written as

$$[\mathcal{L}_{\rm tr}(x), \mathcal{L}_{\rm tr}(x')]\rho = -[A_{\mu}(x), A_{\nu}(x')][j^{\mu}(x) j^{\nu}(x'), \rho]. \tag{16}$$

It is useful to introduce current super-operators $J_+(x)$ and $J_-(x)$ by means of

$$J^{\mu}_{+}(x)\rho \equiv j^{\mu}(x)\rho, \qquad J^{\mu}_{-}(x)\rho \equiv \rho j^{\mu}(x). \tag{17}$$

Thus, $J_+(x)$ is defined to be the current density acting from the left, while $J_-(x)$ acts from the right on an arbitrary density. With the help of these definitions we may write the commutator of the Liouville super-operators as

$$\begin{aligned}[\mathcal{L}_{\rm tr}(x), \mathcal{L}_{\rm tr}(x')] = &- [A_{\mu}(x), A_{\nu}(x')] J^{\mu}_{+}(x) J^{\nu}_{+}(x') \\ &+ [A_{\mu}(x), A_{\nu}(x')] J^{\mu}_{-}(x) J^{\nu}_{-}(x').\end{aligned}$$

Inserting this result into Eq. (13), we can write Eq. (12) as

$$\begin{aligned}\rho_m(t_f) = T^j_{\leftarrow} \Bigg(\exp \Bigg[&\int_{t_i}^{t_f} \mathrm{d}^4x \mathcal{L}_{\rm C}(x) \\ &-\frac{1}{2} \int_{t_i}^{t_f} \mathrm{d}^4x \int_{t_i}^{t_f} \mathrm{d}^4x' \theta(t-t') [A_{\mu}(x), A_{\nu}(x')] J^{\mu}_{+}(x) J^{\nu}_{+}(x') \\ &+\frac{1}{2} \int_{t_i}^{t_f} \mathrm{d}^4x \int_{t_i}^{t_f} \mathrm{d}^4x' \theta(t-t') [A_{\mu}(x), A_{\nu}(x')] J^{\mu}_{-}(x) J^{\nu}_{-}(x') \Bigg] \\ &\cdot \mathrm{tr}_f \left\{ \exp \left[\int_{t_i}^{t_f} \mathrm{d}^4x \mathcal{L}_{\rm tr}(x) \right] \rho(t_i) \right\} \Bigg).\end{aligned} \tag{18}$$

This is an exact formal representation for the reduced density matrix of the matter variables. Note that the time-ordering of the radiation degrees of freedom has been removed and that they enter Eq. (18) only through the functional

$$W[J_+, J_-] \equiv \mathrm{tr}_f \left\{ \exp \left[\int_{t_i}^{t_f} \mathrm{d}^4x \mathcal{L}_{\rm tr}(x) \rho(t_i) \right] \right\}, \tag{19}$$

since the commutator of the A-fields is a c-number function.

3 The Influence Phase Functional of QED

The functional (19) involves an average over the field variables with respect to the initial state $\rho(t_i)$ of the combined matter-field system. It therefore contains all correlations in the initial state of the total system. Here, we are interested in the destruction of coherence. Our central goal is thus to investigate how correlations are built up through the interaction between matter and radiation field. We therefore consider now an initial state of low entropy which is given by a product state of the form

$$\rho(t_i) = \rho_m(t_i) \otimes \rho_f, \tag{20}$$

where $\rho_m(t_i)$ is the density matrix of the matter at the initial time and the density matrix of the radiation field describes an equilibrium state at temperature T,

$$\rho_f = \frac{1}{Z_f} \exp(-\beta H_f). \tag{21}$$

Here, H_f denotes the Hamiltonian of the free radiation field and the quantity $Z_f = \mathrm{tr}_f[\exp(-\beta H_f)]$ is the partition function with $\beta = 1/k_B T$. In the following we shall denote by

$$\langle \mathcal{O} \rangle_f \equiv \mathrm{tr}_f \{\mathcal{O}\rho_f\} \tag{22}$$

the average of some quantity $\mathcal{O}$ with respect to the thermal equilibrium state (21).

The influence of the special choice (20) for the initial condition can be eliminated by pushing $t_i \to -\infty$ and by switching on the interaction adiabatically. This is the usual procedure used in Quantum Field Theory in order to define asymptotic states and the S-matrix. The matter and the field variables are then described as *in*-fields, obeying free field equations with renormalized mass. These fields generate physical one-particle states from the interacting ground state.

For an arbitrary initial condition $\rho(t_i)$ the functional $W[J_+, J_-]$ can be determined, for example, by means of a cumulant expansion. Since the initial state (20) is Gaussian with respect to the field variables and since the Liouville super-operator $\mathcal{L}_{\mathrm{tr}}(x)$ is linear in the radiation field, the cumulant expansion terminates after the second order term. In addition, a linear term does not appear in the expansion because of $\langle A_\mu(x) \rangle_f = 0$. Thus we immediately obtain

$$W[J_+, J_-] = \exp\left[\frac{1}{2}\int_{t_i}^{t_f} \mathrm{d}^4x \int_{t_i}^{t_f} \mathrm{d}^4x' \langle \mathcal{L}_{\mathrm{tr}}(x)\mathcal{L}_{\mathrm{tr}}(x')\rangle_f\right] \rho_m(t_i). \tag{23}$$

Inserting the definition for the Liouville super-operator $\mathcal{L}_{\mathrm{tr}}(x)$ into the exponent of this expression one finds after some algebra,

$$\begin{aligned}
&\frac{1}{2}\int_{t_i}^{t_f} \mathrm{d}^4x \int_{t_i}^{t_f} \mathrm{d}^4x' \langle \mathcal{L}_{\mathrm{tr}}(x)\mathcal{L}_{\mathrm{tr}}(x')\rangle_f \rho_m \\
&\equiv -\frac{1}{2}\int_{t_i}^{t_f} \mathrm{d}^4x \int_{t_i}^{t_f} \mathrm{d}^4x' \mathrm{tr}_f \left\{[\mathcal{H}_{\mathrm{tr}}(x), [\mathcal{H}_{\mathrm{tr}}(x'), \rho_m \otimes \rho_f]]\right\}
\end{aligned}$$

$$= -\frac{1}{2}\int_{t_i}^{t_f} \mathrm{d}^4x \int_{t_i}^{t_f} \mathrm{d}^4x' \, \big[\langle A_\nu(x')A_\mu(x)\rangle_f J_+^\mu(x)J_+^\nu(x') \\ +\langle A_\mu(x)A_\nu(x')\rangle_f J_-^\mu(x)J_-^\nu(x') \\ -\langle A_\nu(x')A_\mu(x)\rangle_f J_+^\mu(x)J_-^\nu(x') \\ -\langle A_\mu(x)A_\nu(x')\rangle_f J_-^\mu(x)J_-^\nu(x')\big]\, \rho_m.$$

On using this result together with Eq. (23), Eq. (18) can be cast into the form,

$$\rho_m(t_f) = T_{\leftarrow}^j \Big(\exp\Big[\int_{t_i}^{t_f} \mathrm{d}^4x \mathcal{L}_C(x) \\ +\frac{1}{2}\int_{t_i}^{t_f} \mathrm{d}^4x \int_{t_i}^{t_f} \mathrm{d}^4x' \{ -\big(\theta(t-t')[A_\mu(x), A_\nu(x')] \\ +\langle A_\nu(x')A_\mu(x)\rangle_f\big) J_+^\mu(x)J_+^\nu(x') \\ +\big(\theta(t-t')[A_\mu(x), A_\nu(x')] \\ -\langle A_\mu(x)A_\nu(x')\rangle_f\big) J_-^\mu(x)J_-^\nu(x') \\ +\langle A_\nu(x')A_\mu(x)\rangle_f J_+^\mu(x)J_-^\nu(x') \\ +\langle A_\mu(x)A_\nu(x')\rangle_f J_-^\mu(x)J_+^\nu(x')\}\Big]\Big)\rho_m(t_i). \tag{24}$$

At this stage it is useful to introduce a new notation for the correlation functions of the electromagnetic field, namely the Feynman propagator and its complex conjugated ($T_{\rightarrow}$ denotes the anti-chronological time-ordering),

$$\mathrm{i}D_F(x-x')_{\mu\nu} \equiv \langle T_{\leftarrow}(A_\mu(x)A_\nu(x'))\rangle_f \\ = \theta(t-t')[A_\mu(x), A_\nu(x')] + \langle A_\nu(x')A_\mu(x)\rangle_f, \\ \mathrm{i}D_F^*(x-x')_{\mu\nu} \equiv -\langle T_{\rightarrow}(A_\mu(x)A_\nu(x'))\rangle_f \\ = \theta(t-t')[A_\mu(x), A_\nu(x')] - \langle A_\mu(x)A_\nu(x')\rangle_f, \tag{25}$$

as well as the two-point correlation functions

$$D_+(x-x')_{\mu\nu} \equiv \langle A_\mu(x)A_\nu(x')\rangle_f, \\ D_-(x-x')_{\mu\nu} \equiv \langle A_\nu(x')A_\mu(x)\rangle_f. \tag{26}$$

As is easily verified these functions are related through

$$-\mathrm{i}D_F(x-x')_{\mu\nu} + \mathrm{i}D_F^*(x-x')_{\mu\nu} + D_+(x-x')_{\mu\nu} + D_-(x-x')_{\mu\nu} = 0. \tag{27}$$

With the help of this notation the density matrix of the matter can now be written as follows,

$$\rho_m(t_f) = T_{\leftarrow}^j \Big(\exp\Big[\int_{t_i}^{t_f} \mathrm{d}^4x \mathcal{L}_c(x) \tag{28}$$

$$+\frac{1}{2}\int_{t_i}^{t_f} \mathrm{d}^4x \int_{t_i}^{t_f} \mathrm{d}^4x' \{-\mathrm{i}D_F(x-x')_{\mu\nu} J_+^\mu(x) J_+^\nu(x')$$
$$+\mathrm{i}D_F^*(x-x')_{\mu\nu} J_-^\mu(x) J_-^\nu(x')$$
$$+D_-(x-x')_{\mu\nu} J_+^\mu(x) J_-^\nu(x')$$
$$+D_+(x-x')_{\mu\nu} J_-^\mu(x) J_+^\nu(x')\}]) \, \rho_m(t_i).$$

This equation provides an exact representation for the matter density matrix which takes on the desired form: It involves the electromagnetic field variables only through the various two-point correlation functions introduced above. One observes that the dynamics of the matter variables is given by a time-ordered exponential function whose exponent is a bilinear functional of the current super-operators $J_\pm(x)$. Formally we may write Eq. (28) as

$$\rho_m(t_f) = T_{\leftarrow}^{j} \exp\left(\mathrm{i}\Phi[J_+, J_-]\right) \rho_m(t_i), \tag{29}$$

where we have introduced an *influence phase functional*

$$\begin{aligned} \mathrm{i}\Phi[J_+, J_-] = \int_{t_i}^{t_f} \mathrm{d}^4x \mathcal{L}_C(x) + \frac{1}{2}\int_{t_i}^{t_f} \mathrm{d}^4x \int_{t_i}^{t_f} \mathrm{d}^4x' \\ \times \left\{-\mathrm{i}D_F(x-x')_{\mu\nu} J_+^\mu(x) J_+^\nu(x') + \mathrm{i}D_F^*(x-x')_{\mu\nu} J_-^\mu(x) J_-^\nu(x') \right. \\ \left. +D_-(x-x')_{\mu\nu} J_+^\mu(x) J_-^\nu(x') + D_+(x-x')_{\mu\nu} J_-^\mu(x) J_+^\nu(x')\right\}. \end{aligned} \tag{30}$$

It should be remarked that the influence phase $\Phi[J_+, J_-]$ is both a functional of the quantities $J_\pm(x)$ and a super-operator which acts in the space of density matrices of the matter degrees of freedom. There are several alternative methods which could be used to arrive at an expression of the form (30) as, for example, path integral techniques [1] or Schwinger's closed time-path method [11]. The expression (30) for the influence phase functional has been given in Ref. [12] without the Coulomb term and for the special case of zero temperature. In our derivation we have combined super-operator techniques with methods from field theory, which seems to be the most direct way to obtain a representation of the reduced density matrix.

For the study of decoherence phenomena another equivalent formula for the influence phase functional will be useful. To this end we define the commutator function

$$\begin{aligned} D(x-x')_{\mu\nu} &\equiv \mathrm{i}[A_\mu(x), A_\nu(x')] \\ &= \mathrm{i}\left(D_+(x-x')_{\mu\nu} - D_-(x-x')_{\mu\nu}\right) \end{aligned} \tag{31}$$

and the anti-commutator function

$$\begin{aligned} D_1(x-x')_{\mu\nu} &\equiv \langle\{A_\mu(x), A_\nu(x')\}\rangle_f \\ &= D_+(x-x')_{\mu\nu} + D_-(x-x')_{\mu\nu}. \end{aligned} \tag{32}$$

Of course, the previously introduced correlation functions may be expressed in terms of $D(x-x')_{\mu\nu}$ and $D_1(x-x')_{\mu\nu}$,

$$D_+(x-x')_{\mu\nu} = \frac{1}{2}D_1(x-x')_{\mu\nu} - \frac{\mathrm{i}}{2}D(x-x')_{\mu\nu}, \quad (33)$$

$$D_-(x-x')_{\mu\nu} = \frac{1}{2}D_1(x-x')_{\mu\nu} + \frac{\mathrm{i}}{2}D(x-x')_{\mu\nu}, \quad (34)$$

$$iD_F(x-x')_{\mu\nu} = \frac{1}{2}D_1(x-x')_{\mu\nu} - \frac{\mathrm{i}}{2}\mathrm{sign}(t-t')D(x-x')_{\mu\nu}, \quad (35)$$

$$-iD_F^*(x-x')_{\mu\nu} = \frac{1}{2}D_1(x-x')_{\mu\nu} + \frac{\mathrm{i}}{2}\mathrm{sign}(t-t')D(x-x')_{\mu\nu}. \quad (36)$$

Correspondingly, we define a commutator super-operator $J_c(x)$ and an anti-commutator super-operator $J_a(x)$ by means of

$$J_c^\mu(x)\rho \equiv [j^\mu(x),\rho], \qquad J_a^\mu(x)\rho \equiv \{j^\mu(x),\rho\}, \quad (37)$$

which are related to the previously introduced super-operators $J_\pm^\mu(x)$ by

$$J_c^\mu(x) = J_+^\mu(x) - J_-^\mu(x), \qquad J_a^\mu(x) = J_+^\mu(x) + J_-^\mu(x). \quad (38)$$

In terms of these quantities the influence phase functional may now be written as

$$\mathrm{i}\Phi[J_c,J_a] = \int_{t_i}^{t_f} \mathrm{d}^4x\mathcal{L}_C(x) + \int_{t_i}^{t_f}\mathrm{d}^4x\int_{t_i}^{t}\mathrm{d}^4x'\left\{\frac{\mathrm{i}}{2}D(x-x')_{\mu\nu}J_c^\mu(x)J_a^\nu(x') - \frac{1}{2}D_1(x-x')_{\mu\nu}J_c^\mu(x)J_c^\nu(x')\right\}. \quad (39)$$

This form of the influence phase functional will be particularly useful later on. It represents the influence of the radiation field on the matter dynamics in terms of the two fundamental 2-point correlation functions $D(x-x')$ and $D_1(x-x')$. Note that the double space-time integral in Eq. (39) is already a time-ordered integral since the integration over $t' = x'^0$ extends over the time interval from t_i to $t = x^0$.

For a physical discussion of these results it may be instructive to compare Eq. (28) with the structure of a Markovian quantum master equation in Lindblad form [3],

$$\frac{d\rho_m}{dt} = -\mathrm{i}[H_m,\rho_m] + \sum_i\left(A_i\rho_m A_i^\dagger - \frac{1}{2}A_i^\dagger A_i\rho_m - \frac{1}{2}\rho_m A_i^\dagger A_i\right), \quad (40)$$

where H_m generates the coherent evolution and the A_i denote a set of operators, the Lindblad operators, labeled by some index i. One observes that the terms of the influence phase functional involving the current super-operators in the combinations J_+J_- and J_-J_+ correspond to the gain terms in the Lindblad equation having the form $A_i \rho_m A_i^\dagger$. These terms may be interpreted as describing the back action on the reduced system of the matter degrees of freedom induced by "real" processes in which photons are absorbed or emitted. The presence of these terms leads to a transformation of pure states into statistical mixtures. Namely, if we disregard the terms containing the combinations J_+J_- and J_-J_+ the remaining expression takes the form

$$\rho_m(t_f) \approx U(t_f, t_i)\rho_m(t_i)U^\dagger(t_f, t_i), \tag{41}$$

where

$$U(t_f, t_i) = T_{\leftarrow}^{j} \exp\left[-\mathrm{i}\int_{t_i}^{t_f} \mathrm{d}^4x \mathcal{H}_C(x) \right. \tag{42}$$
$$\left. -\frac{\mathrm{i}}{2}\int_{t_i}^{t_f} \mathrm{d}^4x \int_{t_i}^{t_f} \mathrm{d}^4x' D_F(x-x')_{\mu\nu} j^\mu(x) j^\nu(x')\right].$$

Eq. (41) shows that the contributions involving the Feynman propagators and the combinations J_+J_+ and J_-J_- of super-operators preserve the purity of states [12]. Recall that all correlations functions have been defined in terms of the transversal radiation field. We may turn to the covariant form of the correlation functions if we replace at the same time the current density by its transversal component j_{tr}^μ. The expression (42) is then seen to contain the vacuum-to-vacuum amplitude $A[j]$ of the electromagnetic field in the presence of a classical, transversal current density $j_{\mathrm{tr}}^\mu(x)$ [13],

$$A[j] = \exp\left[-\frac{\mathrm{i}}{2}\int \mathrm{d}^4x \int \mathrm{d}^4x' D_F(x-x')_{\mu\nu} j_{\mathrm{tr}}^\mu(x) j_{\mathrm{tr}}^\nu(x')\right]. \tag{43}$$

With the help of the decomposition (35) of the Feynman propagator into a real and an imaginary part we find

$$A[j] = \exp\left[\mathrm{i}\left(S^{(1)} + \mathrm{i}S^{(2)}\right)\right]. \tag{44}$$

The vacuum-to-vacuum amplitude is thus represented in terms of a complex action functional with the real part

$$S^{(1)} = \frac{1}{4}\int \mathrm{d}^4x \int \mathrm{d}^4x' \mathrm{sign}(t-t') D(x-x')_{\mu\nu} j_{\mathrm{tr}}^\mu(x) j_{\mathrm{tr}}^\nu(x'), \tag{45}$$

and with the imaginary part

$$S^{(2)} = \frac{1}{4}\int \mathrm{d}^4x \int \mathrm{d}^4x' D_1(x-x')_{\mu\nu} j_{\mathrm{tr}}^\mu(x) j_{\mathrm{tr}}^\nu(x'). \tag{46}$$

The imaginary part $S^{(2)}$ yields the probability that no photon is emitted by the current j_{tr}^μ,

$$|A[j]|^2 = \exp\left(-2S^{(2)}\right). \tag{47}$$

In covariant form we have

$$D(x-x')_{\mu\nu} = -\frac{1}{2\pi}\text{sign}(t-t')\delta[(x-x')^2]g_{\mu\nu}, \tag{48}$$

and, hence,

$$S^{(1)} = -\frac{1}{8\pi}\int \mathrm{d}^4x \int \mathrm{d}^4x' \delta[(x-x')^2] j^{\text{tr}}_{\mu}(x) j^{\mu}_{\text{tr}}(x'). \tag{49}$$

This is the classical Feynman-Wheeler action. It describes the classical motion of a system of charged particles by means of a non-local action which arises after the elimination of the degrees of freedom of the electromagnetic radiation field. In the following sections we will demonstrate that it is just the imaginary part $S^{(2)}$ which leads to the destruction of coherence of the matter degrees of freedom.

4 The Interaction of a Single Electron with the Radiation Field

In this section we shall apply the foregoing general theory to the case of a single electron interacting with the radiation field where we confine ourselves to the non-relativistic approximation. It will be seen that this simple case already contains the basic physical mechanism leading to decoherence.

4.1 Representation of the Electron Density Matrix in the Non-Relativistic Approximation

The starting point will be the representation (29) for the reduced matter density with expression (39) for the influence phase functional Φ. It must be remembered that the correlation functions $D(x-x')_{\mu\nu}$ and $D_1(x-x')_{\mu\nu}$ have been defined in terms of the transversal radiation field using Coulomb gauge and that they thus involve projections onto the transversal component. In fact, we have the replacements,

$$D(x-x')_{\mu\nu} \longrightarrow D(x-x')_{ij} = -\left(\delta_{ij} - \frac{\partial_i\partial_j}{\Delta}\right) D(x-x')$$

for the commutator functions, and

$$D_1(x-x')_{\mu\nu} \longrightarrow D_1(x-x')_{ij} = +\left(\delta_{ij} - \frac{\partial_i\partial_j}{\Delta}\right) D_1(x-x')$$

for the anti-commutator function, where

$$D(x-x') = -\mathrm{i}\int \frac{\mathrm{d}^3k}{2(2\pi)^3\omega}\left[\exp\left(-\mathrm{i}k(x-x')\right) - \exp\left(\mathrm{i}k(x-x')\right)\right],$$

and

$$D_1(x-x') = \int \frac{\mathrm{d}^3k}{2(2\pi)^3\omega} \left[\exp\left(-\mathrm{i}k(x-x')\right) + \exp\left(\mathrm{i}k(x-x')\right)\right] \coth\left(\beta\omega/2\right),$$

with the notation $k^\mu = (\omega, \boldsymbol{k}) = (|\boldsymbol{k}|, \boldsymbol{k})$ for the components of the wave vector. It should be noted that the commutator function is independent of the temperature, while the anti-commutator function does depend on T through the factor $\coth(\beta\omega/2) = 1 + 2N(\omega)$, where $N(\omega)$ is the average number of photons in a mode with frequency ω. Hence, invoking the non-relativistic (dipole) approximation we may replace

$$D(x-x')_{ij} \longrightarrow D(t-t')_{ij} = \delta_{ij} D(t-t') = \delta_{ij} \int_0^\infty \mathrm{d}\omega J(\omega) \sin\omega(t-t'), \quad (50)$$

and

$$\begin{aligned} D_1(x-x')_{ij} \longrightarrow D_1(t-t')_{ij} &= \delta_{ij} D_1(t-t') \\ &= \delta_{ij} \int_0^\infty \mathrm{d}\omega J(\omega) \coth\left(\beta\omega/2\right) \cos\omega(t-t'), \end{aligned} \quad (51)$$

where we have introduced the spectral density

$$J(\omega) = \frac{e^2}{3\pi^2} \omega \Theta(\Omega - \omega), \quad (52)$$

with some ultraviolet cutoff Ω (see below). It is important to stress here that the spectral density increases with the first power of the frequency ω. Had we used dipole coupling $-e\boldsymbol{x}\cdot\boldsymbol{E}$ of the electron coordinate $\boldsymbol{x}$ to the electric field strength $\boldsymbol{E}$, the corresponding spectral density would be proportional to the third power of the frequency. This means that the coupling to the radiation field in the dipole approximation may be described as a special case of the famous Caldeira-Leggett model [2] and that in the language of the theory of quantum Brownian motion [14] the radiation field constitutes a super-Ohmic environment [15,16]. Note also that we now include the factor e^2 into the definition of the correlation function. Within the non-relativistic approximation we may thus replace the current density by

$$\boldsymbol{j}(t, \boldsymbol{x}) \longrightarrow \frac{\boldsymbol{p}(t)}{2m} \delta(\boldsymbol{x} - \boldsymbol{x}(t)) + \delta(\boldsymbol{x} - \boldsymbol{x}(t)) \frac{\boldsymbol{p}(t)}{2m}, \quad (53)$$

where $\boldsymbol{p}(t)$ and $\boldsymbol{x}(t)$ denote the momentum and position operator of the electron in the interaction picture with respect to the Hamiltonian

$$H_m = \frac{\boldsymbol{p}^2}{2m} + V(\boldsymbol{x}) \quad (54)$$

for the electron, $V(\boldsymbol{x})$ being some external potential.

We are thus led to the following non-relativistic approximation of Eq. (29),

$$\rho_m(t_f) = T_{\leftarrow}\left(\exp\left[\int_{t_i}^{t_f} dt \int_{t_i}^{t} dt' \left\{\frac{i}{2}D(t-t')\frac{\boldsymbol{p}_c(t)}{m}\frac{\boldsymbol{p}_a(t')}{m} - \frac{1}{2}D_1(t-t')\frac{\boldsymbol{p}_c(t)}{m}\frac{\boldsymbol{p}_c(t')}{m}\right\}\right]\right)\rho_m(t_i). \quad (55)$$

This equation represents the density matrix (neglecting the spin degree of freedom) for a single electron interacting with the radiation field at temperature T. In accordance with the definitions (37) and (38) $\boldsymbol{p}_c$ is a commutator super-operator and $\boldsymbol{p}_a$ an anti-commutator super-operator. In the theory of quantum Brownian motion the function $D(t-t')$ is called the *dissipation kernel*, whereas $D_1(t-t')$ is referred to as *noise kernel.*

4.2 The Path Integral Representation

The reduced density matrix given in Eq. (55) admits an equivalent path integral representation [14] which may be written as follows,

$$\rho_m(\boldsymbol{x}_f, \boldsymbol{x}'_f, t_f) = \int d^3x_i \int d^3x'_i J(\boldsymbol{x}_f, \boldsymbol{x}'_f, t_f; \boldsymbol{x}_i, \boldsymbol{x}'_i, t_i)\rho_m(\boldsymbol{x}_i, \boldsymbol{x}'_i, t_i), \quad (56)$$

with the *propagator function*

$$J(\boldsymbol{x}_f, \boldsymbol{x}'_f, t_f; \boldsymbol{x}_i, \boldsymbol{x}'_i, t_i) = \int D\boldsymbol{x}D\boldsymbol{x}' \exp\left\{i\left(S_m[\boldsymbol{x}] - S_m[\boldsymbol{x}']\right) + i\Phi[\boldsymbol{x}, \boldsymbol{x}']\right\}. \quad (57)$$

This is a double path integral which is to be extended over all paths $\boldsymbol{x}(t)$ and $\boldsymbol{x}'(t)$ with the boundary conditions

$$\boldsymbol{x}'(t_i) = \boldsymbol{x}'_i, \qquad \boldsymbol{x}'(t_f) = \boldsymbol{x}'_f, \qquad \boldsymbol{x}(t_i) = \boldsymbol{x}_i, \qquad \boldsymbol{x}(t_f) = \boldsymbol{x}_f. \quad (58)$$

$S_m[\boldsymbol{x}]$ denotes the action functional for the electron,

$$S_m[\boldsymbol{x}] = \int_{t_i}^{t_f} dt \left(\frac{1}{2}m\dot{\boldsymbol{x}}^2 - V(\boldsymbol{x})\right), \quad (59)$$

while the influence phase functional becomes,

$$i\Phi[\boldsymbol{x}, \boldsymbol{x}'] = \int_{t_i}^{t_f} dt \int_{t_i}^{t} dt' \left\{\frac{i}{2}D(t-t')\left(\dot{\boldsymbol{x}}(t) - \dot{\boldsymbol{x}}'(t)\right)\left(\dot{\boldsymbol{x}}(t') + \dot{\boldsymbol{x}}'(t')\right) - \frac{1}{2}D_1(t-t')\left(\dot{\boldsymbol{x}}(t) - \dot{\boldsymbol{x}}'(t)\right)\left(\dot{\boldsymbol{x}}(t') - \dot{\boldsymbol{x}}'(t')\right)\right\}. \quad (60)$$

We define the new variables

$$\boldsymbol{q} = \boldsymbol{x} - \boldsymbol{x}', \qquad \boldsymbol{r} = \frac{1}{2}(\boldsymbol{x} + \boldsymbol{x}'), \quad (61)$$

and set, for simplicity, the initial time equal to zero, $t_i = 0$. We may then write Eq. (56) as

$$\rho_m(\boldsymbol{r}_f, \boldsymbol{q}_f, t_f) = \int \mathrm{d}^3 r_i \int \mathrm{d}^3 q_i J(\boldsymbol{r}_f, \boldsymbol{q}_f, t_f; \boldsymbol{r}_i, \boldsymbol{q}_i) \rho_m(\boldsymbol{r}_i, \boldsymbol{q}_i, 0). \tag{62}$$

The propagator function

$$J(\boldsymbol{r}_f, \boldsymbol{q}_f, t_f; \boldsymbol{r}_i, \boldsymbol{q}_i) = \int \mathrm{D}\boldsymbol{r} \int \mathrm{D}\boldsymbol{q} \exp\{\mathrm{i}\mathcal{A}[\boldsymbol{r}, \boldsymbol{q}]\} \tag{63}$$

is a double path integral over all path $\boldsymbol{r}(t)$, $\boldsymbol{q}(t)$ satisfying the boundary conditions,

$$\boldsymbol{r}(0) = \boldsymbol{r}_i, \qquad \boldsymbol{r}(t_f) = \boldsymbol{r}_f, \qquad \boldsymbol{q}(0) = \boldsymbol{q}_i, \qquad \boldsymbol{q}(t_f) = \boldsymbol{q}_f. \tag{64}$$

The weight factor for the paths $\boldsymbol{r}(t)$, $\boldsymbol{q}(t)$ is defined in terms of an effective action $\mathcal{A}$ functional,

$$\begin{aligned} \mathcal{A}[\boldsymbol{r}, \boldsymbol{q}] = & \int_0^{t_f} \mathrm{d}t \left(m\dot{\boldsymbol{r}}\dot{\boldsymbol{q}} - V(\boldsymbol{r} + \frac{1}{2}\boldsymbol{q}) + V(\boldsymbol{r} - \frac{1}{2}\boldsymbol{q}) \right) \\ & + \int_0^{t_f} \mathrm{d}t \int_0^{t_f} \mathrm{d}t' \theta(t - t') D(t - t') \dot{\boldsymbol{q}}(t) \dot{\boldsymbol{r}}(t') \\ & + \frac{\mathrm{i}}{4} \int_0^{t_f} \mathrm{d}t \int_0^{t_f} \mathrm{d}t' D_1(t - t') \dot{\boldsymbol{q}}(t) \dot{\boldsymbol{q}}(t'). \end{aligned} \tag{65}$$

The first variation of $\mathcal{A}$ is found to be

$$\begin{aligned} \delta\mathcal{A} = - \int_0^{t_f} \mathrm{d}t \Bigg\{ & \delta\boldsymbol{q}(t) \Bigg[m\ddot{\boldsymbol{r}}(t) + \frac{1}{2}\boldsymbol{\nabla}_{\boldsymbol{r}}(V(\boldsymbol{r} + \frac{1}{2}\boldsymbol{q}) + V(\boldsymbol{r} - \frac{1}{2}\boldsymbol{q})) \\ & + \frac{\mathrm{d}}{\mathrm{d}t} \int_0^t \mathrm{d}t' D(t - t') \dot{\boldsymbol{r}}(t') + \frac{\mathrm{i}}{2} \frac{\mathrm{d}}{\mathrm{d}t} \int_0^{t_f} \mathrm{d}t' D_1(t - t') \dot{\boldsymbol{q}}(t') \Bigg] \\ & + \delta\boldsymbol{r}(t) \Bigg[m\ddot{\boldsymbol{q}}(t) + 2\boldsymbol{\nabla}_{\boldsymbol{q}}(V(\boldsymbol{r} + \frac{1}{2}\boldsymbol{q}) + V(\boldsymbol{r} - \frac{1}{2}\boldsymbol{q})) \\ & + \frac{\mathrm{d}}{\mathrm{d}t} \int_t^{t_f} \mathrm{d}t' D(t' - t) \dot{\boldsymbol{q}}(t') \Bigg] \Bigg\}, \end{aligned} \tag{66}$$

which leads to the classical equations of motion,

$$\begin{aligned} & m\ddot{\boldsymbol{r}}(t) + \frac{1}{2}\boldsymbol{\nabla}_{\boldsymbol{r}}(V(\boldsymbol{r} + \frac{1}{2}\boldsymbol{q}) + V(\boldsymbol{r} - \frac{1}{2}\boldsymbol{q})) + \frac{\mathrm{d}}{\mathrm{d}t} \int_0^t \mathrm{d}t' D(t - t') \dot{\boldsymbol{r}}(t') \\ & = -\frac{\mathrm{i}}{2} \frac{\mathrm{d}}{\mathrm{d}t} \int_0^{t_f} \mathrm{d}t' D_1(t - t') \dot{\boldsymbol{q}}(t'), \end{aligned} \tag{67}$$

and

$$m\ddot{\boldsymbol{q}}(t) + 2\boldsymbol{\nabla}_{\boldsymbol{q}}(V(\boldsymbol{r} + \frac{1}{2}\boldsymbol{q}) + V(\boldsymbol{r} - \frac{1}{2}\boldsymbol{q})) + \frac{\mathrm{d}}{\mathrm{d}t} \int_t^{t_f} \mathrm{d}t' D(t' - t) \dot{\boldsymbol{q}}(t') = 0. \tag{68}$$

4.3 The Abraham-Lorentz Equation

The real part of the equation of motion (67), which is obtained by setting the right-hand side equal to zero, yields the famous Abraham-Lorentz equation for the electron [17]. It describes the radiation damping through the damping kernel $D(t-t')$ [15]. To see this we write the real part of Eq. (67) as

$$m\ddot{\boldsymbol{r}}(t) + \frac{\mathrm{d}}{\mathrm{d}t}\int_0^t \mathrm{d}t' D(t-t')\dot{\boldsymbol{r}}(t') = \boldsymbol{F}_{\mathrm{ext}}(t), \tag{69}$$

where $\boldsymbol{F}_{\mathrm{ext}}(t)$ denotes an external force derived from the potential V. The damping kernel can be written (see Eqs. (50) and (52))

$$\begin{aligned} D(t-t') &= \int_0^\Omega \mathrm{d}\omega \frac{e^2}{3\pi^2}\omega \sin\omega(t-t') = \frac{e^2}{3\pi^2}\frac{\mathrm{d}}{\mathrm{d}t'}\int_0^\Omega \mathrm{d}\omega \cos\omega(t-t') \\ &= \frac{e^2}{3\pi^2}\frac{\mathrm{d}}{\mathrm{d}t'}\frac{\sin\Omega(t-t')}{t-t'} \equiv \frac{e^2}{3\pi^2}\frac{\mathrm{d}}{\mathrm{d}t'}f(t-t'), \end{aligned}$$

where we have introduced the function

$$f(t) \equiv \frac{\sin\Omega t}{t}. \tag{70}$$

To be specific the UV-cutoff Ω is taken to be

$$\hbar\Omega = mc^2, \tag{71}$$

which implies that

$$\Omega = \frac{mc^2}{\hbar} = \frac{c}{\bar{\lambda}_C}, \tag{72}$$

where

$$\bar{\lambda}_C = \frac{\hbar}{mc} \tag{73}$$

is the Compton wavelength. For an electron we have

$$\bar{\lambda}_C \approx 3.8 \times 10^{-13}\mathrm{m} \qquad \text{and} \qquad \Omega \approx 0.78 \times 10^{21}\mathrm{s}^{-1}. \tag{74}$$

The term of the equation of motion (69) involving the damping kernel can be written as follows,

$$\begin{aligned} \frac{\mathrm{d}}{\mathrm{d}t}\int_0^t \mathrm{d}t' D(t-t')\dot{\boldsymbol{r}}(t') &= \frac{e^2}{3\pi^2}\frac{\mathrm{d}}{\mathrm{d}t}\int_0^t \mathrm{d}t' \left[\frac{\mathrm{d}}{\mathrm{d}t'}f(t-t')\right]\dot{\boldsymbol{r}}(t') \\ &= \frac{e^2}{3\pi^2}\frac{\mathrm{d}}{\mathrm{d}t}\left[-\int_0^t \mathrm{d}t' f(t-t')\ddot{\boldsymbol{r}}(t') + f(0)\dot{\boldsymbol{r}}(t) - f(t)\dot{\boldsymbol{r}}(0).\right] \end{aligned} \tag{75}$$

For times t such that $\Omega t \gg 1$, i.e. $t \gg 10^{-21}$s, we may replace

$$f(t) \longrightarrow \pi\delta(t), \tag{76}$$

and approximate $f(t) \approx 0$, while Eq. (70) yields $f(0) = \Omega$. Thus we obtain,

$$\frac{d}{dt} \int_0^t dt' D(t-t') \dot{\boldsymbol{r}}(t') = \frac{e^2}{3\pi^2} \frac{d}{dt} \left[-\frac{\pi}{2} \ddot{\boldsymbol{r}}(t) + \Omega \dot{\boldsymbol{r}}(t) \right], \tag{77}$$

which finally leads to the equation of motion,

$$\left(m + \frac{e^2 \Omega}{3\pi^2} \right) \dot{\boldsymbol{v}}(t) - \frac{e^2}{6\pi} \ddot{\boldsymbol{v}}(t) = \boldsymbol{F}_{\text{ext}}(t), \tag{78}$$

where $\boldsymbol{v} = \dot{\boldsymbol{r}}$ is the velocity. This is the famous Abraham-Lorentz equation [17]. The term proportional to the third derivative of $\boldsymbol{r}(t)$ describes the damping of the electron motion through the emitted radiation. This term does not depend on the cutoff frequency, while the cutoff-dependent term yields a renormalization of the electron mass,

$$m_R = m + \Delta m = m + \frac{e^2 \Omega}{3\pi^2}. \tag{79}$$

It is important to note that the electro-magnetic mass Δm diverges linearly with the cutoff. The equation of motion (78) can be obtained heuristically by means of the Larmor formula for the power radiated by an accelerated charge. More rigorously, it has been derived by Abraham and by Lorentz from the conservation law for the field momentum, assuming a spherically symmetric charge distribution and that the momentum is of purely electromagnetic origin [17].

For the cutoff Ω chosen above we get

$$\Delta m = \frac{me^2}{3\pi^2} = \frac{4}{3\pi} \alpha m, \tag{80}$$

and, hence,

$$\frac{\Delta m}{m} = \frac{4}{3\pi} \alpha \approx 0.0031.$$

The decomposition (79) of the mass is, however, unphysical, since the electron is never observed without its self-field and the associated field momentum. In other words, we have to identify the renormalized mass m_R with the observed physical mass which enables us to write Eq. (78) as

$$m_R \left[\dot{\boldsymbol{v}}(t) - \tau_0 \ddot{\boldsymbol{v}}(t) \right] = \boldsymbol{F}_{\text{ext}}(t). \tag{81}$$

Here, the radiation damping term has been written in terms of a characteristic radiation time scale τ_0 given by

$$\tau_0 \equiv \frac{e^2}{6\pi m_R} = \frac{2}{3} r_e \approx 0.6 \times 10^{-23} \text{s}, \tag{82}$$

where r_e denotes the classical electron radius,

$$r_e = \frac{e^2}{4\pi m_R} = \alpha \bar{\lambda}_C \approx 2.8 \times 10^{-15} \text{m}. \tag{83}$$

It is well-known that Eq. (81), being a classical equation of motion for the electron, leads to the problem of exponentially increasing *runaway solutions.* Namely, for $\boldsymbol{F}_{\text{ext}} = 0$ we have

$$\dot{\boldsymbol{v}} - \tau_0 \ddot{\boldsymbol{v}} = 0. \tag{84}$$

In addition to the trivial solution of a constant velocity, $\boldsymbol{v} = \text{const}$, one also finds the solution

$$\dot{\boldsymbol{v}}(t) = \dot{\boldsymbol{v}}(0) \exp(t/\tau_0),$$

describing an exponential growth of the acceleration for $\dot{\boldsymbol{v}}(0) \neq 0$. In order to exclude these solutions one imposes the boundary condition

$$\dot{\boldsymbol{v}}(t) \longrightarrow 0 \qquad \text{for} \qquad t \longrightarrow \infty,$$

if $\boldsymbol{F}_{\text{ext}}$ also vanishes in this limit. This boundary condition can be implemented by rewriting Eq. (81) as an integro-differential equation

$$m_R \ddot{\boldsymbol{r}}(t) = \int_0^\infty \mathrm{d}s \exp(-s) \boldsymbol{F}_{\text{ext}}(t + \tau_0 s). \tag{85}$$

On differentiating Eq. (85) with respect to time, it is easily verified that one is led back to Eq. (81). However, for $\boldsymbol{F}_{\text{ext}} = 0$ it follows immediately from Eq. (85) that $\boldsymbol{v} = \text{const}$, such that runaway solutions are excluded.

On the other hand, Eq. (85) shows that the acceleration depends upon the future value of the force. Hence, the electron reacts to signals lying a time of order τ_0 in the future, which is the phenomenon of pre-acceleration. This phenomenon should, however, not be taken too seriously, since the description is only classical. The time scale τ_0 corresponds to a length scale r_e which is smaller than the Compton wavelength $\bar{\lambda}_C$ by a factor of α, such that a quantum mechanical treatment of the problem is required.

4.4 Construction of the Decoherence Function

In this subsection we derive the explicit form of the propagator function (63) for the reduced electron density matrix in the case of quadratic potentials,

$$V(\boldsymbol{x}) = \frac{1}{2} m_R \omega_0^2 \boldsymbol{x}^2. \tag{86}$$

Our aim is to introduce and to determine the *decoherence function* which provides a quantitative measure for the degree of decoherence. On using

$$\begin{aligned} V(\boldsymbol{r} + \boldsymbol{q}/2) + V(\boldsymbol{r} - \boldsymbol{q}/2) &= m_R \omega_0^2 \boldsymbol{r}^2 + m_R \omega_0^2 \boldsymbol{q}^2/4 \\ -V(\boldsymbol{r} + \boldsymbol{q}/2) + V(\boldsymbol{r} - \boldsymbol{q}/2) &= -m_R \omega_0^2 \boldsymbol{r} \cdot \boldsymbol{q}, \end{aligned} \tag{87}$$

the classical equations of motion take the form

$$m_R \left[\ddot{\boldsymbol{r}}(t) + \omega_0^2 \int_0^\infty \mathrm{d}s \exp(-s) \boldsymbol{r}(t+\tau_0 s) \right] = -\frac{\mathrm{i}}{2} \frac{\mathrm{d}}{\mathrm{d}t} \int_0^{t_f} \mathrm{d}t' D_1(t-t') \dot{\boldsymbol{q}}(t') \quad (88)$$

$$m_R \left[\ddot{\boldsymbol{q}}(t) + \omega_0^2 \int_0^\infty \mathrm{d}s \exp(-s) \boldsymbol{q}(t-\tau_0 s) \right] = 0. \quad (89)$$

Note, that Eq. (89) is the *backward equation* of the real part of Eq. (88). More precisely, if $\boldsymbol{q}(t)$ solves Eq. (89), then $\boldsymbol{r}(t) \equiv \boldsymbol{q}(t_f - t)$ is a solution of (88) with the right-hand side set equal to zero.

The above equations of motion lead to the following renormalized action functional

$$\begin{aligned} \mathcal{A}[\boldsymbol{r},\boldsymbol{q}] = & \int_0^{t_f} \mathrm{d}t \, m_R \left[\dot{\boldsymbol{r}}(t)\dot{\boldsymbol{q}}(t) - \omega_0^2 \boldsymbol{q}(t) \int_0^\infty \mathrm{d}s \exp(-s) \boldsymbol{r}(t+\tau_0 s) \right] \\ & + \frac{\mathrm{i}}{4} \int_0^{t_f} \mathrm{d}t \int_0^{t_f} \mathrm{d}t' D_1(t-t') \dot{\boldsymbol{q}}(t) \dot{\boldsymbol{q}}(t'). \end{aligned} \quad (90)$$

In the following we shall use this renormalised action functional instead of the action given in Eq. (65). By variation with respect to $\boldsymbol{q}(t)$ we immediately obtain Eq. (88), whereas the variation with respect to $\boldsymbol{r}(t)$ yields:

$$-\int_0^{t_f} \mathrm{d}t \, m_R \left[\ddot{\boldsymbol{q}}(t) \delta \boldsymbol{r}(t) + \omega_0^2 \int_0^\infty \mathrm{d}s \exp(-s) \boldsymbol{q}(t) \delta \boldsymbol{r}(t+\tau_0 s) \right] = 0,$$

which implies

$$\begin{aligned} & \int_0^{t_f} \mathrm{d}t \ddot{\boldsymbol{q}}(t) \delta \boldsymbol{r}(t) + \omega_0^2 \int_0^\infty \mathrm{d}s \int_0^{t_f} \mathrm{d}t \exp(-s) \boldsymbol{q}(t) \delta \boldsymbol{r}(t+\tau_0 s) \\ & = \int_0^{t_f} \mathrm{d}t \ddot{\boldsymbol{q}}(t) \delta \boldsymbol{r}(t) + \omega_0^2 \int_0^\infty \mathrm{d}s \int_{\tau_0 s}^{t_f+\tau_0 s} \mathrm{d}t \exp(-s) \boldsymbol{q}(t-\tau_0 s) \delta \boldsymbol{r}(t) \\ & = 0. \end{aligned} \quad (91)$$

In the last time integral we may extend the integration over the time interval from 0 to t_f. This is legitimate since τ_0 is the radiation time scale: By setting this variation of the action equal to zero we thus neglect times of the order of the pre-acceleration time, which directly leads to the equation of motion (89).

Since the action functional is quadratic the propagator function can be determined exactly by evaluating the action along the classical solution and by taking into account Gaussian fluctuations around the classical paths. We therefore assume that $\boldsymbol{r}(t)$ and $\boldsymbol{q}(t)$ are solutions of the classical equations of motion (88) and (89) with boundary conditions (64). The effective action along these solutions may be written as

$$\mathcal{A}_{\mathrm{cl}}[\boldsymbol{r},\boldsymbol{q}] = m_R[\dot{\boldsymbol{r}}_f \boldsymbol{q}_f - \dot{\boldsymbol{r}}_i \boldsymbol{q}_i]$$

$$
-\int_0^{t_f} \mathrm{d}t\, m_R \boldsymbol{q}(t) \left[\ddot{\boldsymbol{r}}(t) + \omega_0^2 \int_0^\infty \mathrm{d}s \exp(-s) \boldsymbol{r}(t + \tau_0 s) \right]
$$
$$
+\frac{\mathrm{i}}{4} \int_0^{t_f} \mathrm{d}t \int_0^{t_f} \mathrm{d}t' D_1(t-t') \dot{\boldsymbol{q}}(t) \dot{\boldsymbol{q}}(t'), \tag{92}
$$

or, equivalently,

$$
\mathcal{A}_{\mathrm{cl}}[\boldsymbol{r}, \boldsymbol{q}] = m_R[\dot{\boldsymbol{r}}_f \boldsymbol{q}_f - \dot{\boldsymbol{r}}_i \boldsymbol{q}_i] + \frac{\mathrm{i}}{2} \int_0^{t_f} \mathrm{d}t\, \boldsymbol{q}(t) \frac{\mathrm{d}}{\mathrm{d}t} \int_0^{t_f} \mathrm{d}t' D_1(t-t') \dot{\boldsymbol{q}}(t')
$$
$$
+\frac{\mathrm{i}}{4} \int_0^{t_f} \mathrm{d}t \int_0^{t_f} \mathrm{d}t' D_1(t-t') \dot{\boldsymbol{q}}(t) \dot{\boldsymbol{q}}(t'). \tag{93}
$$

Eq. (88) shows that the solution $\boldsymbol{r}(t)$ is, in general, complex due to the coupling to $\boldsymbol{q}(t)$ via the noise kernel $D_1(t-t')$. Consider the decomposition of $\boldsymbol{r}(t)$ into real and imaginary part,

$$
\boldsymbol{r}(t) = \boldsymbol{r}^{(1)}(t) + \mathrm{i}\boldsymbol{r}^{(2)}(t), \tag{94}
$$

where $\boldsymbol{r}^{(1)}$ is a solution of the real part of Eq. (88), while $\boldsymbol{r}^{(2)}$ solves its imaginary part,

$$
m_R \left[\ddot{\boldsymbol{r}}^{(2)}(t) + \omega_0^2 \int_0^\infty \mathrm{d}s \exp(-s) \boldsymbol{r}^{(2)}(t + \tau_0 s) \right] = -\frac{1}{2} \frac{\mathrm{d}}{\mathrm{d}t} \int_0^{t_f} \mathrm{d}t' D_1(t-t') \dot{\boldsymbol{q}}(t'). \tag{95}
$$

We now demonstrate that, in order to determine the action along the classical paths, it suffices to find the homogeneous solution $\boldsymbol{r}^{(1)}$ and to insert it in the action functional [14]. In other words we have

$$
\mathcal{A}_{\mathrm{cl}}[\boldsymbol{r}^{(1)}, \boldsymbol{q}] = \mathcal{A}_{\mathrm{cl}}[\boldsymbol{r}, \boldsymbol{q}], \tag{96}
$$

where

$$
\mathcal{A}_{\mathrm{cl}}[\boldsymbol{r}^{(1)}, \boldsymbol{q}] = m_R[\dot{\boldsymbol{r}}_f^{(1)} \boldsymbol{q}_f - \dot{\boldsymbol{r}}_i^{(1)} \boldsymbol{q}_i]
$$
$$
+\frac{\mathrm{i}}{4} \int_0^{t_f} \mathrm{d}t \int_0^{t_f} \mathrm{d}t' D_1(t-t') \dot{\boldsymbol{q}}(t) \dot{\boldsymbol{q}}(t'). \tag{97}
$$

To proof this statement we first deduce from Eq. (95) that

$$
\frac{\mathrm{i}}{2} \int_0^{t_f} \mathrm{d}t\, \boldsymbol{q}(t) \frac{\mathrm{d}}{\mathrm{d}t} \int_0^{t_f} \mathrm{d}t' D_1(t-t') \dot{\boldsymbol{q}}(t')
$$
$$
= -\mathrm{i} m_R \int_0^{t_f} \mathrm{d}t\, \boldsymbol{q}(t) \left[\ddot{\boldsymbol{r}}^{(2)}(t) + \omega_0^2 \int_0^\infty \mathrm{d}s \exp(-s) \boldsymbol{r}^{(2)}(t + \tau_0 s) \right]
$$
$$
= -\mathrm{i} m_R [\dot{\boldsymbol{r}}_f^{(2)} \boldsymbol{q}_f - \dot{\boldsymbol{r}}_i^{(2)} \boldsymbol{q}_i]
$$
$$
-\mathrm{i} m_R \int_0^{t_f} \mathrm{d}t \left[\boldsymbol{r}^{(2)}(t) \ddot{\boldsymbol{q}}(t) + \omega_0^2 \int_0^\infty \mathrm{d}s \exp(-s) \boldsymbol{r}^{(2)}(t + \tau_0 s) \boldsymbol{q}(t) \right]. \tag{98}
$$

The term within the square brackets is seen to vanish if one employs Eq. (89) and the same arguments that were used to derive the equation of motion from the variation (91) of the action functional. Furthermore, we made use of $\boldsymbol{r}^{(2)}(0) = \boldsymbol{r}^{(2)}(t_f) = 0$ which means that the real part $\boldsymbol{r}^{(1)}(t)$ of the solution satisfies the given boundary conditions. Hence we find

$$\frac{\mathrm{i}}{2}\int_0^{t_f} \mathrm{d}t\, \boldsymbol{q}(t)\frac{\mathrm{d}}{\mathrm{d}t}\int_0^{t_f} \mathrm{d}t' D_1(t-t')\dot{\boldsymbol{q}}(t') = -\mathrm{i}m_R[\dot{\boldsymbol{r}}_f^{(2)}\boldsymbol{q}_f - \dot{\boldsymbol{r}}_i^{(2)}\boldsymbol{q}_i], \tag{99}$$

from which we finally obtain with the help of (93),

$$\begin{aligned} \mathcal{A}_{\mathrm{cl}}[\boldsymbol{r},\boldsymbol{q}] &= m_R[\dot{\boldsymbol{r}}_f\boldsymbol{q}_f - \dot{\boldsymbol{r}}_i\boldsymbol{q}_i] - \mathrm{i}m_R[\dot{\boldsymbol{r}}_f^{(2)}\boldsymbol{q}_f - \dot{\boldsymbol{r}}_i^{(2)}\boldsymbol{q}_i] \\ &\quad + \frac{\mathrm{i}}{4}\int_0^{t_f}\mathrm{d}t\int_0^{t_f}\mathrm{d}t' D_1(t-t')\dot{\boldsymbol{q}}(t)\dot{\boldsymbol{q}}(t') \\ &= \mathcal{A}_{\mathrm{cl}}[\boldsymbol{r}^{(1)},\boldsymbol{q}]. \end{aligned} \tag{100}$$

This completes the proof of the above statement.

Summarizing, the procedure to determine the propagator function for the electron can now be given as follows. One first solves the equations of motion

$$\ddot{\boldsymbol{r}}(t) + \omega_0^2\int_0^\infty \mathrm{d}s\exp(-s)\boldsymbol{r}(t+\tau_0 s) = 0, \tag{101}$$

$$\ddot{\boldsymbol{q}}(t) + \omega_0^2\int_0^\infty \mathrm{d}s\exp(-s)\boldsymbol{q}(t-\tau_0 s) = 0, \tag{102}$$

together with the boundary conditions (64). With the help of these solutions one then evaluates the classical action,

$$\mathcal{A}_{\mathrm{cl}}[\boldsymbol{r},\boldsymbol{q}] = m_R[\dot{\boldsymbol{r}}_f\boldsymbol{q}_f - \dot{\boldsymbol{r}}_i\boldsymbol{q}_i] + \frac{\mathrm{i}}{4}\int_0^{t_f}\mathrm{d}t\int_0^{t_f}\mathrm{d}t' D_1(t-t')\dot{\boldsymbol{q}}(t)\dot{\boldsymbol{q}}(t'), \tag{103}$$

which immediately yields the propagator function

$$\begin{aligned} J(\boldsymbol{r}_f,\boldsymbol{q}_f,t_f;\boldsymbol{r}_i,\boldsymbol{q}_i) &= N\exp\{\mathrm{i}\mathcal{A}_{\mathrm{cl}}[\boldsymbol{r},\boldsymbol{q}]\} \\ &= N\exp\{\mathrm{i}m_R(\dot{\boldsymbol{r}}_f\boldsymbol{q}_f - \dot{\boldsymbol{r}}_i\boldsymbol{q}_i) + \Gamma(\boldsymbol{q}_f,\boldsymbol{q}_i,t_f)\}. \end{aligned} \tag{104}$$

Here, N is a normalization factor which is determined from the normalization condition

$$\int \mathrm{d}^3 r_f J(\boldsymbol{r}_f,\boldsymbol{q}_f=0,t_f;\boldsymbol{r}_i,\boldsymbol{q}_i) = \delta(\boldsymbol{q}_i). \tag{105}$$

The function $\Gamma(\boldsymbol{q}_f,\boldsymbol{q}_i,t_f)$ introduced in Eq. (104) will be referred to as the *decoherence function*. It is given in terms of the noise kernel $D_1(t-t')$ as

$$\Gamma(\boldsymbol{q}_f,\boldsymbol{q}_i,t_f) = -\frac{1}{4}\int_0^{t_f}\mathrm{d}t\int_0^{t_f}\mathrm{d}t' D_1(t-t')\dot{\boldsymbol{q}}(t)\dot{\boldsymbol{q}}(t'). \tag{106}$$

Explicitly we find with the help of Eq. (51),

$$\Gamma = -\frac{1}{4}\int_0^{t_f} \mathrm{d}t \int_0^{t_f} \mathrm{d}t' \int_0^{\infty} \mathrm{d}\omega J(\omega)\coth(\beta\omega/2)\cos\omega(t-t')\dot{\boldsymbol{q}}(t)\dot{\boldsymbol{q}}(t'). \quad (107)$$

The double time-integral can be written as

$$\mathrm{Re}\int_0^{t_f} \mathrm{d}t \int_0^{t_f} \mathrm{d}t' \exp[\mathrm{i}\omega(t-t')]\dot{\boldsymbol{q}}(t)\dot{\boldsymbol{q}}(t') = \left|\int_0^{t_f} \mathrm{d}t \exp(i\omega t)\dot{\boldsymbol{q}}(t)\right|^2 . \quad (108)$$

Hence, the decoherence function takes the form

$$\Gamma(\boldsymbol{q}_f, \boldsymbol{q}_i, t_f) = -\frac{1}{4}\int_0^{\infty} \mathrm{d}\omega J(\omega)\coth(\beta\omega/2)\left|\boldsymbol{Q}(\omega)\right|^2 , \quad (109)$$

where we have introduced

$$\boldsymbol{Q}(\omega) \equiv \int_0^{t_f} \mathrm{d}t \exp(\mathrm{i}\omega t)\dot{\boldsymbol{q}}(t). \quad (110)$$

It can be seen from the above expressions that Γ is a non-positive function. The decoherence function will be demonstrated below to describe the reduction of electron coherence through the influence of the radiation field.

5 Decoherence Through the Emission of Bremsstrahlung

As an example we shall investigate in this section the most simple case, namely that of a free electron coupled to the radiation field. This case is of particular interest since it allows an exact analytical determination of the decoherence function and already yields a clear physical picture for the decoherence mechanism. Having determined the decoherence function, we proceed with an investigation of its influence on the propagation of electronic wave packets.

5.1 Determination of the Decoherence Function

We set $\omega_0 = 0$ to describe the free electron. The equations of motion (101) and (102) with the boundary conditions (64) can easily be solved to yield

$$\boldsymbol{r}(t) = \boldsymbol{r}_i + \frac{\boldsymbol{r}_f - \boldsymbol{r}_i}{t_f}t, \qquad \boldsymbol{q}(t) = \boldsymbol{q}_i + \frac{\boldsymbol{q}_f - \boldsymbol{q}_i}{t_f}t. \quad (111)$$

Making use of Eq. (104) and determining the normalization factor from Eq. (105) we thus get the propagator function,

$$J(\boldsymbol{r}_f, \boldsymbol{q}_f, t_f; \boldsymbol{r}_i, \boldsymbol{q}_i) = \left(\frac{m_R}{2\pi t_f}\right)^3 \exp\left\{\mathrm{i}\frac{m_R}{t_f}(\boldsymbol{r}_f - \boldsymbol{r}_i)(\boldsymbol{q}_f - \boldsymbol{q}_i) + \Gamma(\boldsymbol{q}_f, \boldsymbol{q}_i, t_f)\right\} . \quad (112)$$

As must have been expected J is invariant under space translations since it depends only on the difference $\boldsymbol{r}_f - \boldsymbol{r}_i$. Furthermore, one easily recognizes that the contribution

$$G(\boldsymbol{r}_f - \boldsymbol{r}_i, \boldsymbol{q}_f - \boldsymbol{q}_i, t_f) \equiv \left(\frac{m_R}{2\pi t_f}\right)^3 \exp\left\{ \mathrm{i}\frac{m_R}{t_f}(\boldsymbol{r}_f - \boldsymbol{r}_i)(\boldsymbol{q}_f - \boldsymbol{q}_i) \right\} \tag{113}$$

is simply the propagator function for the density matrix of a free electron with mass m_R for a vanishing coupling to the radiation field. We can thus write the electron density matrix as follows,

$$\rho_m(\boldsymbol{r}_f, \boldsymbol{q}_f, t_f) = \int \mathrm{d}^3 r_i \int \mathrm{d}^3 q_i G(\boldsymbol{r}_f - \boldsymbol{r}_i, \boldsymbol{q}_f - \boldsymbol{q}_i, t_f) \\ \times \exp\left\{ \Gamma(\boldsymbol{q}_f, \boldsymbol{q}_i, t_f) \right\} \rho_m(\boldsymbol{r}_i, \boldsymbol{q}_i, 0), \tag{114}$$

which exhibits that the decoherence function Γ describes the influence of the radiation field on the electron motion.

We proceed with an explicit calculation of the decoherence function. It follows from Eqs. (110) and (111) that

$$\boldsymbol{Q}(\omega) = \int_0^{t_f} \mathrm{d}t \exp(\mathrm{i}\omega t)\frac{\boldsymbol{q}_f - \boldsymbol{q}_i}{t_f} = \frac{\exp(\mathrm{i}\omega t_f) - 1}{\mathrm{i}\omega}\boldsymbol{w}, \tag{115}$$

where

$$\boldsymbol{w} \equiv \frac{1}{t_f}(\boldsymbol{q}_f - \boldsymbol{q}_i). \tag{116}$$

Therefore, the decoherence function is found to be

$$\Gamma = -\frac{e^2 \boldsymbol{w}^2}{6\pi^2} \int_0^{\Omega} \mathrm{d}\omega \frac{1 - \cos\omega t_f}{\omega} \coth(\beta\omega/2), \tag{117}$$

where we have used expression (52) for the spectral density $J(\omega)$. The decoherence function may be decomposed into a vacuum contribution Γ_{vac} and a thermal contribution Γ_{th},

$$\Gamma = \Gamma_{\mathrm{vac}} + \Gamma_{\mathrm{th}}, \tag{118}$$

where

$$\Gamma_{\mathrm{vac}} = -\frac{e^2 \boldsymbol{w}^2}{6\pi^2} \int_0^{\Omega} \mathrm{d}\omega \frac{1 - \cos\omega t_f}{\omega} \tag{119}$$

and

$$\Gamma_{\mathrm{th}} = -\frac{e^2 \boldsymbol{w}^2}{6\pi^2} \int_0^{\Omega} \mathrm{d}\omega \frac{1 - \cos\omega t_f}{\omega} \left[\coth(\beta\omega/2) - 1\right]. \tag{120}$$

The frequency integral appearing in the vacuum contribution can be evaluated in the following way. Substituting $x = \omega t_f$ we get

$$\int_0^{\Omega} \mathrm{d}\omega \frac{1 - \cos\omega t_f}{\omega} = \int_0^{\Omega t_f} \mathrm{d}x \frac{1 - \cos x}{x} = \ln \Omega t_f + C + O\left(\frac{1}{\Omega t_f}\right), \tag{121}$$

where $C \approx 0.577$ is Euler's constant [18]. For $\Omega t_f \gg 1$ we obtain asymptotically

$$\Gamma_{\rm vac} \approx -\frac{e^2 \boldsymbol{w}^2}{6\pi^2} \ln \Omega t_f = -\frac{e^2}{6\pi^2} \ln \Omega t_f \frac{(\boldsymbol{q}_f - \boldsymbol{q}_i)^2}{t_f^2}. \tag{122}$$

To determine the thermal contribution $\Gamma_{\rm th}$ we first write Eq. (120) as follows,

$$\Gamma_{\rm th} = -\frac{e^2 \boldsymbol{w}^2}{6\pi^2} \int_0^{t_f} {\rm d}t \int_0^{\Omega} {\rm d}\omega \left[\coth(\beta\omega/2) - 1\right] \sin \omega t \equiv -\frac{e^2 \boldsymbol{w}^2}{6\pi^2} I. \tag{123}$$

Introducing the integration variable $x = \beta\omega$ we can cast the double integral I into the form

$$I = \frac{1}{\beta} \int_0^{t_f} {\rm d}t \int_0^{\beta\Omega} {\rm d}x \left[\coth(x/2) - 1\right] \sin\left(tx/\beta\right).$$

Here, we have $\beta\Omega = \hbar\Omega / k_B T$ and, using the cutoff $\hbar\Omega = mc^2$, we get

$$\beta\Omega = \frac{mc^2}{k_B T}.$$

For temperatures T obeying

$$k_B T \ll mc^2 \tag{124}$$

the upper limit of the x-integral may be shifted from $\beta\Omega$ to ∞. Condition (124) states that

$$\frac{\hbar^2}{m k_B T} \gg \frac{\hbar^2}{m^2 c^2},$$

which means that the thermal wavelength $\bar{\lambda}_{\rm th} = \hbar / \sqrt{2 m k_B T}$ is much larger than the Compton wavelength,

$$\bar{\lambda}_{\rm th} \gg \bar{\lambda}_C. \tag{125}$$

Thermal and Compton wavelength are of equal size at a temperature of about 10^9 Kelvin. Condition (124) therefore means that $T \ll 10^9$ K. Under this condition we now obtain

$$\begin{aligned} I &\approx \frac{1}{\beta} \int_0^{t_f} {\rm d}t \int_0^{\infty} {\rm d}x \left[\coth(x/2) - 1\right] \sin\left(tx/\beta\right) \\ &= \frac{1}{\beta} \int_0^{t_f} {\rm d}t \left[\pi \coth\left(\frac{\pi t}{\beta}\right) - \frac{\beta}{t}\right] \\ &= \ln\left(\frac{\sinh\left(\pi t_f/\beta\right)}{\pi t_f/\beta}\right), \end{aligned} \tag{126}$$

where we have employed the formula

$$\int_0^\infty \mathrm{d}x \, [\coth(x/2) - 1] \sin \tau x = \pi \coth(\pi\tau) - \frac{1}{\tau}. \tag{127}$$

The quantity

$$\tau_B \equiv \frac{\beta}{\pi} = \frac{\hbar}{\pi k_B T} \approx 2.4 \cdot 10^{-12} \text{ s/T[K]} \tag{128}$$

represents the correlation time of the thermal radiation field. Putting these results together we get the following expression for the thermal contribution to the decoherence function,

$$\Gamma_{\text{th}} \approx -\frac{e^2}{6\pi^2} \ln\left(\frac{\sinh(t_f/\tau_B)}{t_f/\tau_B}\right) \frac{(\boldsymbol{q}_f - \boldsymbol{q}_i)^2}{t_f^2}. \tag{129}$$

Adding this expression to the vacuum contribution (122) and introducing $\alpha = e^2/4\pi\hbar c$ and further factors of c, we can finally write the expression for the decoherence function as

$$\Gamma(\boldsymbol{q}_f, \boldsymbol{q}_i, t_f) \approx -\frac{2\alpha}{3\pi}\left[\ln \Omega t_f + \ln\left(\frac{\sinh(t_f/\tau_B)}{t_f/\tau_B}\right)\right] \frac{(\boldsymbol{q}_f - \boldsymbol{q}_i)^2}{(ct_f)^2}. \tag{130}$$

Alternatively, we may write

$$\Gamma(\boldsymbol{q}_f, \boldsymbol{q}_i, t_f) = -\frac{(\boldsymbol{q}_f - \boldsymbol{q}_i)^2}{2L(t_f)^2}, \tag{131}$$

where the quantity $L(t_f)$ defined by

$$L(t_f)^2 \equiv \frac{3\pi}{4\alpha}\left[\ln \Omega t_f + \ln\left(\frac{\sinh(t_f/\tau_B)}{t_f/\tau_B}\right)\right]^{-1} \cdot (ct_f)^2 \tag{132}$$

may be interpreted as a time-dependent *coherence length.*

The vacuum contribution Γ_{vac} to the decoherence function (130) apparently diverges with the logarithm of the cutoff Ω. This is, however, an artificial divergence which can be seen as follows. The decoherence function is defined in terms of the Fourier transform $\boldsymbol{Q}(\omega)$ of $\dot{\boldsymbol{q}}(t)$, see Eqs. (109) and (110). Evaluating $\boldsymbol{Q}(\omega)$ as in Eq. (115) we assume that the velocity is zero prior to the initial time $t = 0$, that it suddenly jumps to the value given by Eq. (116), and that it again jumps to zero at time t_f. This implies a force having the shape of two δ-function pulses around $t = 0$ and $t = t_f$. Such a force acts over two infinitely small time intervals and leads to sharp edges in the classical path. More realistically one has to consider a finite time scale τ_p for the action of the force which must be still large compared to the radiation time scale τ_0. We may interpret the time scale τ_p as a *preparation time* since it represents the time required to prepare the initial state of a moving electron.

A natural, physical cutoff frequency of the order $\Omega \sim 1/\tau_p$ is thus introduced by the preparation time scale τ_p and we may set

$$\Omega t_f = \frac{t_f}{\tau_p} \tag{133}$$

in the following. It should be noted that the weak logarithmic dependence on Ω shows that the precise value of the preparation time scale τ_p is rather irrelevant. The important point is that the preparation time introduces a new time scale which removes the dependence on the cutoff. The vacuum decoherence function can thus be written,

$$\Gamma_{\text{vac}} \approx -\frac{2\alpha}{3\pi} \ln\left(\frac{t_f}{\tau_p}\right) \frac{(\boldsymbol{q}_f - \boldsymbol{q}_i)^2}{(ct_f)^2}, \tag{134}$$

showing that it vanishes for large times essentially as t_f^{-2}.

The thermal contribution Γ_{th} is determined by the thermal correlation time τ_B. For $T \longrightarrow 0$ we have $\tau_B \longrightarrow \infty$, and this contribution vanishes. For large times $t_f \gg \tau_B$ the thermal decoherence function may be approximated by

$$\Gamma_{\text{th}} \approx -\frac{2\alpha}{3\pi} \frac{t_f}{\tau_B} \frac{(\boldsymbol{q}_f - \boldsymbol{q}_i)^2}{(ct_f)^2}, \tag{135}$$

which shows that Γ_{th} vanishes as t_f^{-1}. Thus, for short times the vacuum contribution dominates, whereas the thermal contribution is dominant for large times. Both contributions Γ_{vac} and Γ_{th} are plotted separately in Fig. 1 which clearly shows the crossover between the two regions of time.

Eq. (132) implies that the vacuum coherence length is roughly of the order

$$L(t_f)_{\text{vac}} \sim c \cdot t_f. \tag{136}$$

To see this let us assume a typical preparation time scale of the order $\tau_p \sim 10^{-21}$s. If we take t_f to be of the order of 1s we find that $\ln(t_f/\tau_p) \sim 48$. In the rather extreme case $t_f \sim 10^{17}$, which is of the order of the age of the universe, we get $\ln(t_f/\tau_p) \sim 87$. On using $3\pi/4\alpha \approx 322$ and Eq. (132) for $T = 0$ one is led to the estimate (136).

5.2 Wave Packet Propagation

Having obtained an expression for the decoherence function Γ we now proceed with a detailed discussion of its physical significance. For this purpose it will be helpful to investigate first how Γ affects the time-evolution of an electronic wave packet. We consider the initial wave function at time $t = 0$,

$$\psi_0(\boldsymbol{x}) = \left(\frac{1}{2\pi\sigma_0^2}\right)^{3/4} \exp\left[-\frac{(\boldsymbol{x}-\boldsymbol{a})^2}{4\sigma_0^2} - \mathrm{i}\boldsymbol{k}_0(\boldsymbol{x}-\boldsymbol{a})\right], \tag{137}$$

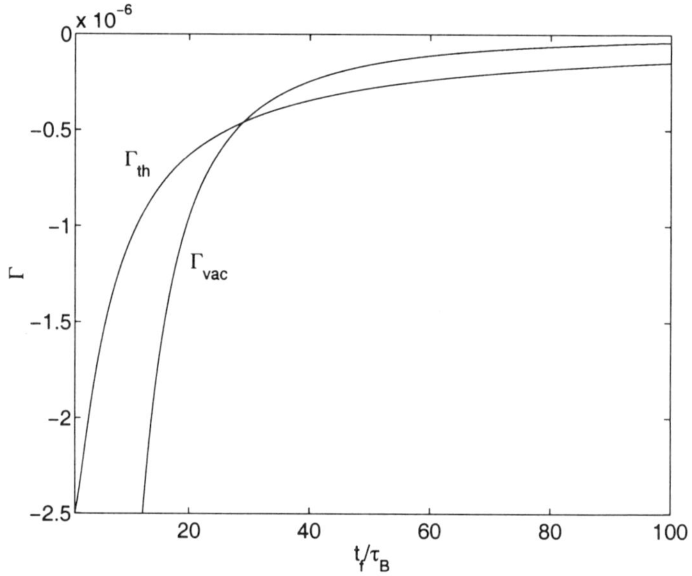

Fig. 1. The vacuum contribution $\Gamma_{\rm vac}$ and the thermal contribution $\Gamma_{\rm th}$ of the decoherence function Γ (Eq. (130)). For a fixed value $|\boldsymbol{q}_f - \boldsymbol{q}_i| = 0.1 \cdot c\tau_B$, the two contributions are plotted against the time t_f which is measured in units of the thermal correlation time τ_B. The temperature was chosen to be $T = 1$K. One observes the decrease of both contributions for increasing time, demonstrating the vanishing of decoherence effects for long times. The thermal contribution $\Gamma_{\rm th}$ vanishes as t_f^{-1}, while the vacuum contribution $\Gamma_{\rm vac}$ decays essentially as t_f^{-2}, leading to a crossover between two regimes dominated by the vacuum and by the thermal contribution, respectively.

describing a Gaussian wave packet centered at $\boldsymbol{x} = \boldsymbol{a}$ with width σ_0. With the help of Eqs. (113), (114) and (131) we get the position space probability density at the final time t_f,

$$\begin{aligned} \rho_m(\boldsymbol{r}_f, t_f) &\equiv \rho_m(\boldsymbol{r}_f, \boldsymbol{q}_f = 0, t_f) \\ &= \int \mathrm{d}^3 r_i \int \mathrm{d}^3 q_i \left(\frac{m_R}{2\pi t_f}\right)^3 \exp\left[-\frac{i m_R}{t_f}(\boldsymbol{r}_f - \boldsymbol{r}_i)\boldsymbol{q}_i - \frac{\boldsymbol{q}_i^2}{2L(t_f)^2}\right] \\ &\qquad \times \psi_0(\boldsymbol{r}_i + \frac{1}{2}\boldsymbol{q}_i)\psi_0^*(\boldsymbol{r}_i - \frac{1}{2}\boldsymbol{q}_i). \end{aligned} \tag{138}$$

The Gaussian integrals may easily be evaluated with the result,

$$\rho_m(\boldsymbol{r}_f, t_f) = \left(\frac{1}{2\pi\sigma(t_f)^2}\right)^{3/2} \exp\left[-\frac{(\boldsymbol{r}_f - \boldsymbol{b})^2}{2\sigma(t_f)^2}\right], \tag{139}$$

where

$$\boldsymbol{b} \equiv \boldsymbol{a} - \frac{\boldsymbol{k}_0 t_f}{m_R} \tag{140}$$

and

$$\sigma(t_f)^2 \equiv \sigma_0^2 + \frac{t_f^2}{4m_R^2\sigma_0^2} + \frac{t_f^2}{m_R^2 L^2}. \tag{141}$$

This shows that the wave packet propagates very much like that of a free Schrödinger particle with physical mass m_R. The centre $\boldsymbol{b}$ of the probability density moves with velocity $-\boldsymbol{k}_0/m_R$, while its spreading, given by Eq. (141), is similar to the spreading $\sigma(t_f)^2_{\rm free}$ which is obtained from the free Schrödinger equation,

$$\sigma(t_f)^2_{\rm free} = \sigma_0^2 + \frac{t_f^2}{4m_R^2\sigma_0^2}. \tag{142}$$

If we write

$$\sigma(t_f)^2 = \sigma_0^2 + \frac{t_f^2}{4m_R^2\sigma_0^2}\left(1 + \frac{4\sigma_0^2}{L(t_f)^2}\right) \tag{143}$$

we observe that the decoherence function affects the probability density only though the width $\sigma(t_f)$ and leads to an increase of the spreading. In view of the estimate (136) the correction term in Eq. (143) is, however, small for times satisfying

$$L(t_f) \sim c \cdot t_f \gg \sigma_0. \tag{144}$$

This means that the influence of the radiation field can safely be neglected for times which are large compared to the time it takes a light signal to travel the width of the wave packet.

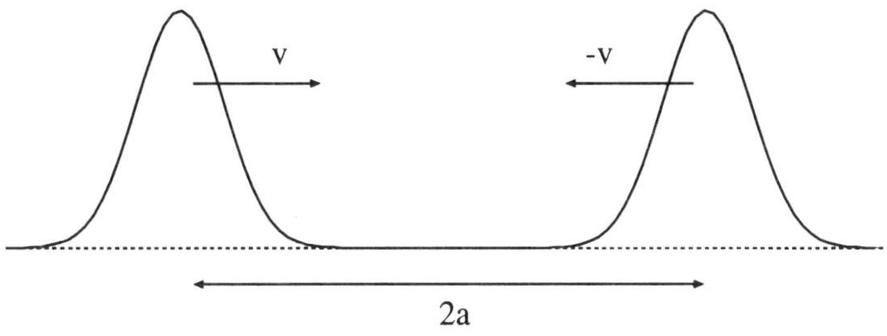

Fig. 2. Sketch of the interference experiment used to determine the decoherence factor. Two Gaussian wave packets with initial separation $2a$ approach each other with opposite velocities of equal magnitude $v = k_0/m_R$.

Let us now study the evolution of a superposition of two Gaussian wave packets separated by a distance $2a$. This case has been studied already by

Barone and Caldeira [15] who find, however, a different result. We assume that the packets have equal widths σ_0 and that they are centered initially at $\boldsymbol{x} = \pm\boldsymbol{a} = \pm(a,0,0)$. The packets are supposed to approach each other with the speed $v = k_0/m_R > 0$ (see Fig. 2). For simplicity the motion is assumed to occur along the x-axis. Thus we have the initial state

$$\psi_0(\boldsymbol{x}) = A_1 \left(\frac{1}{2\pi\sigma_0^2}\right)^{3/4} \exp\left[-\frac{(\boldsymbol{x}-\boldsymbol{a})^2}{4\sigma_0^2} - \mathrm{i}\boldsymbol{k}_0(\boldsymbol{x}-\boldsymbol{a})\right] + A_2 \left(\frac{1}{2\pi\sigma_0^2}\right)^{3/4} \exp\left[-\frac{(\boldsymbol{x}+\boldsymbol{a})^2}{4\sigma_0^2} + \mathrm{i}\boldsymbol{k}_0(\boldsymbol{x}+\boldsymbol{a})\right], \tag{145}$$

where $\boldsymbol{k}_0 = (k_0, 0, 0)$ and A_1, A_2 are complex amplitudes. Our aim is to determine the interference pattern that arises in the moment of collision of the two packets at $\boldsymbol{x} = 0$. Using again Eqs. (113), (114) and (131) and doing the Gaussian integrals we find

$$\rho_m(\boldsymbol{r}_f, t_f) = \left(\frac{1}{2\pi\sigma(t_f)^2}\right)^{3/2} \exp\left[-\frac{\boldsymbol{r}_f^2}{2\sigma(t_f)^2}\right] \times \left\{|A_1|^2 + |A_2|^2 + 2\mathrm{Re}A_1 A_2^* \exp[\varphi(\boldsymbol{r}_f)]\right\}. \tag{146}$$

We recognize a Gaussian envelope centered at $\boldsymbol{r}_f = 0$ with width $\sigma(t_f)$, an incoherent sum $|A_1|^2 + |A_2|^2$, and an interference term proportional to $A_1 A_2^*$. The interference term involves a complex phase given by

$$\varphi(\boldsymbol{r}_f) = -2\mathrm{i}\boldsymbol{k}_0\boldsymbol{r}_f(1-\varepsilon) - \frac{2a^2}{L(t_f)^2}(1-\varepsilon). \tag{147}$$

The term $-2\mathrm{i}\boldsymbol{k}_0\boldsymbol{r}_f$ describes the usual interference pattern as it occurs for a free Schrödinger particle, while the contribution $2\mathrm{i}\boldsymbol{k}_0\boldsymbol{r}_f\varepsilon$ leads to a modification of the period of the pattern. The final time $t_f = am_R/k_0$ is the collision time and

$$v = \frac{a}{t_f} = \frac{k_0}{m_R} \tag{148}$$

is the speed of the wave packets. The factor ε is given by

$$\varepsilon \equiv \frac{t_f^2}{m_R^2 L(t_f)^2 \sigma(t_f)^2} = \left(1 + \frac{L(t_f)^2}{4\sigma_0^2} + \frac{m_R^2\sigma_0^2 L(t_f)^2}{t_f^2}\right)^{-1}. \tag{149}$$

Obviously we always have $0 < \varepsilon < 1$. Furthermore, for the situation considered in Eq. (144) we have $\epsilon \ll 1$. Thus, we get

$$\varphi(\boldsymbol{r}_f) = -2\mathrm{i}\boldsymbol{k}_0\boldsymbol{r}_f - \frac{2a^2}{L(t_f)^2}. \tag{150}$$

The last expression clearly reveals that the real part of the phase $\varphi(\boldsymbol{r}_f)$ describes decoherence, namely a reduction of the interference contrast described by the factor

$$D = \exp\left[-\frac{(2a)^2}{2L(t_f)^2}\right] = \exp\left[-\frac{\text{distance}^2}{2(\text{coherence length})^2}\right], \tag{151}$$

which multiplies the interference term. As was to be expected from the general formula for Γ, the decoherence factor D is determined by the ratio of the distance of the two wave packets to the coherence length.

Alternatively, we can write the decoherence factor in terms of the velocity (148) of the wave packets. In the vacuum case we then get

$$D_{\text{vac}} = \exp\left[-\frac{8\alpha}{3\pi}\ln\left(\frac{t_f}{\tau_p}\right)\left(\frac{v}{c}\right)^2\right]. \tag{152}$$

This clearly demonstrates that it is the motion of the wave packets which is responsible for the reduction of of the interference contrast: If one sets into relative motion the two components of the superposition in order to check locally their capability to interfere, a decoherence effect is caused by the creation of a radiation field. As can be seen from Eq. (117) the spectrum of the radiation field emitted through the moving charge is proportional to $1/\omega$ which is a typical signature for the emission of bremsstrahlung. Thus we observe that the physical origin for the loss of coherence described by the decoherence function is the creation of bremsstrahlung.

It is important to recognize that the frequency integral of Eq. (117) converges for $\omega \to 0$, see Eq. (121). The decoherence function Γ is thus infrared convergent which is obviously due to the fact that we consider here a process on a finite time scale t_f. This means that we have a natural infrared cutoff of the order of $\Omega_{\text{min}} \sim 1/t_f$, in addition to the natural ultraviolet cutoff $\Omega \sim 1/\tau_p$ introduced earlier. The important conclusion is that the decoherence function is therefore infrared as well as ultraviolet convergent.

It might be instructive, finally, to compare our results with the corresponding expressions which are derived from the famous Caldeira-Leggett master equation in the high-temperature limit (see, e.g. [3]). From the latter one finds the following expression for the coherence length

$$L(t_f)^2_{\text{CL}} = \frac{\bar{\lambda}^2_{\text{th}}}{2\gamma t_f}, \tag{153}$$

where γ is the relaxation rate. This is to be compared with the expressions (132) for the coherence length. For large temperatures we have the following dominant time and temperature dependence,

$$L(t_f)^2 \sim \frac{t_f}{T} \quad \text{and} \quad L(t_f)^2_{\text{CL}} \sim \frac{1}{Tt_f}. \tag{154}$$

Hence, while both expressions for the coherence length are proportional to the inverse temperature, the time dependence is completely different. Namely, for $t_f \longrightarrow \infty$ we have

$$L(t_f)^2 \longrightarrow \infty \qquad \text{and} \qquad L(t_f)^2_{\mathrm{CL}} \longrightarrow 0, \tag{155}$$

and, therefore, complete coherence in the case of bremsstrahlung and total destruction of coherence in the Caldeira-Leggett case.

6 The Harmonically Bound Electron in the Radiation Field

As a further illustration let us investigate briefly the case of an electron in the radiation field moving in a harmonic external potential. Another approach to this problem may be found in [19], where the authors arrive, however, at the conclusion that there is no decoherence effect in the vacuum case.

We take $\omega_0 > 0$ and solve the equation of motion (101) with the help of the ansatz

$$r(t) = r_0 \exp(zt), \tag{156}$$

where, for simplicity, we consider the motion to be one-dimensional. Substituting this ansatz into (101) one is led to a cubic equation for z,

$$z^2 - \tau_0 z^3 + \omega_0^2 = 0. \tag{157}$$

For vanishing coupling to the radiation field ($\tau_0 = 0$) the solutions are located at $z_\pm = \pm i\omega_0$, describing the free motion of a harmonic oscillator with frequency ω_0.

For $\tau_0 > 0$ the cubic equation has three roots, one is real and the other two are complex conjugated to each other. The real root corresponds to the runaway solution and must be discarded. Let us assume that the period of the oscillator is large compared to the radiation time,

$$\tau_0 \ll \frac{1}{\omega_0}. \tag{158}$$

Because of $\tau_0 \sim 10^{-24}$s this assumption is well satisfied even in the regime of optical frequencies. We may thus determine the complex roots to lowest order in $\omega_0\tau_0$,

$$z_\pm = \pm i\omega_0 - \frac{1}{2}\tau_0\omega_0^2. \tag{159}$$

The purely imaginary roots $\pm i\omega_0$ of the undisturbed harmonic oscillator are thus shifted into the negative half plane under the influence of the radiation field. The negative real part describes the radiative damping. In fact, we see that $r(t)$ decays as $\exp(-\gamma t/2)$, where

$$\gamma = \tau_0\omega_0^2 = \frac{2}{3}\alpha\frac{\hbar\omega_0^2}{m_R c^2} \tag{160}$$

is the damping constant for radiation damping [17]. In the following we consider times t_f of the order of magnitude of one period $\omega_0 t_f \sim 1$. Because of $\gamma t_f = (\omega_0\tau_0)(\omega_0 t_f)$ we then have $\gamma t_f \sim \tau_0\omega_0 \ll 1$. In this case the damping can be neglected and we may use the free solution in order to determine the decoherence function.

Let us consider again the case of a superposition of two Gaussian wave packets in the harmonic potential. The packets are initially separated by a distance $2a$ and approach each other with opposite velocities of equal magnitude such that they collide after a quarter of a period, $t_f = \pi/2\omega_0$. The corresponding free solution $q(t)$ is therefore given by

$$q(t) = q_i \cos\omega_0 t + q_f \sin\omega_0 t. \tag{161}$$

To describe the situation we have in mind we take $q_i = 2a$ (initial separation of the wave packets) and $q_f = 0$ (to get the probability density). Hence, we have

$$\dot{q}(t) = -2a\omega_0 \sin\omega_0 t, \tag{162}$$

and we evaluate the Fourier transform,

$$\begin{aligned} Q(\omega) &= \int_0^{t_f} \mathrm{d}t \exp(\mathrm{i}\omega t)\dot{q}(t) \\ &= a\omega_0 \left[\frac{\exp(\mathrm{i}[\omega+\omega_0]t_f)-1}{\omega+\omega_0} - \frac{\exp(\mathrm{i}[\omega-\omega_0]t_f)-1}{\omega-\omega_0}\right]. \end{aligned}$$

This yields the decoherence function

$$\begin{aligned} \Gamma &\equiv \Gamma(q_f=0, q_i, t_f) \\ &= -\frac{e^2(a\omega_0)^2}{6\pi^2}\int_0^{\Omega} \mathrm{d}\omega\omega\left[\frac{1-\cos(\omega+\omega_0)t_f}{(\omega+\omega_0)^2} + \frac{1-\cos(\omega-\omega_0)t_f}{(\omega-\omega_0)^2}\right]\coth\left(\frac{\beta\omega}{2}\right). \end{aligned} \tag{163}$$

We discuss the case of zero temperature. The frequency integral in Eq. (163) then approaches asymptotically the value $2\ln\Omega t_f$ which leads to the following expression for the decoherence factor,

$$D_{\mathrm{vac}} = \exp\Gamma_{\mathrm{vac}} = \exp\left[-\frac{8\alpha}{3\pi}\ln\left(\frac{t_f}{\tau_p}\right)\left\langle\left(\frac{v}{c}\right)^2\right\rangle\right]. \tag{164}$$

The interesting point to note here is that this equation is the same as Eq. (152) for the free electron, with the only difference that the square $(v/c)^2$ of the velocity, which was constant in the previous case, must now be replaced with its time averaged value $\langle(v/c)^2\rangle$.

7 Destruction of Coherence of Many-Particle States

For a single electron the vacuum decoherence factor (152) turns out to be very close to 1, as can be illustrated by means of the following numerical example.

We take τ_p to be of the order of 10^{-21}s and t_f of the order of 1s. Using a velocity v which is already as large as 1/10 of the speed of light, one finds that $\Gamma_{\text{vac}} \sim 10^{-2}$, corresponding to a reduction of the interference contrast of about 1%. This demonstrates that the electromagnetic field vacuum is quite ineffective in destroying the coherence of single electrons.

For a superposition of many-particle states the above picture can lead, however, to a dramatic increase of the decoherence effect. Consider the superposition

$$|\psi\rangle = |\psi_1\rangle + |\psi_2\rangle \tag{165}$$

of two well-localized, spatially separated N-particle states $|\psi_1\rangle$ and $|\psi_2\rangle$. We have seen that decoherence results from the imaginary part of the influence phase functional $\Phi[J_c, J_a]$, that is from the last term on the right-hand side of Eq. (39) involving the anti-commutator function $D_1(x-x')_{\mu\nu}$ of the electromagnetic field. Thus, it is the functional

$$\Gamma[J_c] = -\frac{1}{4}\int_{t_i}^{t_f} \mathrm{d}^4x \int_{t_i}^{t_f} \mathrm{d}^4x' D_1(x-x')_{\mu\nu} J_c^{\mu}(x) J_c^{\nu}(x'), \tag{166}$$

which is responsible for decoherence. This shows that the decoherence function for N-electron states scales with the square N^2 of the particle number. Thus we conclude that for the case of the superposition (165) the decoherence function must be multiplied by a factor of N^2, that is the decoherence factor for N-particle states takes the form,

$$D_{\text{vac}}^{N} \sim \exp\left[-\frac{8\alpha}{3\pi}\ln\left(\frac{t_f}{\tau_p}\right)\left(\frac{v}{c}\right)^2 N^2\right]. \tag{167}$$

This scaling with the particle number obviously leads to a dramatic increase of decoherence for the superposition of N-particle states. To give an example we take $N = 6\cdot 10^{23}$, corresponding to 1 mol, and ask for the maximal velocity v leading to a 1% suppression of interference. With the help of (167) we find that $v \sim 10^{-16}$m/s. This means that, in order to perform an interference experiment with 1 mol electrons with only 1% decoherence, a velocity of at most 10^{-16}m/s may be used. For a distance of 1m this implies, for example, that the experiment would take 3×10^8 years!

8 Conclusions

In this paper the equations governing a basic decoherence mechanism occurring in QED have been developed, namely the suppression of coherence through the emission of bremsstrahlung. The latter is created whenever two spatially separated wave packets of a coherent superposition are moved to one place, which is indispensable if one intends to check locally their capability to interfere. We have seen that the decoherence effect through the electromagnetic radiation field is extremely small for single, non-relativistic electrons.

The decoherence mechanism is thus very ineffective on the Compton length scale. An important conclusion is that decoherence does *not* lead to a localization of the particle on arbitrarily small length scales and that no problems with associated UV-divergences arise here.

The decoherence mechanism through bremsstrahlung exhibits a highly non-Markovian character. As a result the usual picture of decoherence as a decay of the off-diagonals in the reduced density matrix does not apply. In fact, consider a superposition of two wave packets with zero velocity. The expression (131) for the decoherence function together with the estimate $L(t_f)_{\rm vac} \sim c \cdot t_f$ for the vacuum coherence length $L(t_f)_{\rm vac}$ show that decoherence effects are negligible for times t_f which are large in comparison to the time it takes light to travel the distance between the wave packets. The off-diagonal terms of the reduced density matrix for the electron do therefore not decay at all, which shows the profound difference between the decoherence mechanism through bremsstrahlung and other decoherence mechanisms (see, e. g. [20]).

A result of particular interest from a fundamental point of view is that coherence can already be destroyed by the presence of the electromagnetic field vacuum if superpositions of many-particle states are considered. An important conclusion which can be drawn from this picture of decoherence in QED refers to various alternative approaches to decoherence and the closely related measurement problem of quantum mechanics: In recent years several attempts have been made to modify the Schrödinger equation by the addition of stochastic terms with the aim to explain the non-existence of macroscopic superpositions through some kind of macrorealism. Namely, the random terms in the Schrödinger equation lead to a spontaneous destruction of superpositions in such a way that macroscopic objects are practically always in definite localized states. Such approaches obviously require the introduction of previously unknown physical constants. In the stochastic theory of Ghirardi, Pearle and Rimini [21], for example, a single particle microscopic jump rate of about $10^{-16}{\rm s}^{-1}$ has to be introduced such that decoherence is extremely weak for single particles but acts sufficiently strong for many particle assemblies. It is interesting to observe that the decoherence effect caused by the presence of the quantum field vacuum yields a similar time scale in a completely natural way without the introduction of new physical parameters. Thus, QED indeed provides a consistent picture of decoherence and it seems unnecessary to propose new ad hoc theories for this purpose.

It must be emphasized that the above picture of decoherence in QED has been derived from the well-established basic postulates of quantum mechanics and quantum field theory. It therefore does not, of course, constitute a logical disprove of alternative approaches. However, it does represent an example for a basic decoherence mechanism in a microscopic quantum field theory. In particular, it provides a unified explanation of decoherence which does not suffer from problems with renormalization (as they occur, e.g. in alternative theories [22]) and which does not exclude a priori the existence of macroscopic quantum coherence. Only under certain well-defined conditions regarding time scales, relative velocities and the structure of the state vector, it is true that decoherence becomes important. Thus, decoherence is traced back to a dynamical effect and not to a modification of the basic principles of quantum mechanics.

In this paper we have discussed in detail only the non-relativistic approximation of the reduced electron dynamics. For a treatment of the full relativistic theory, including a Lorentz invariant characterization of the decoherence induced by the vacuum field, one can start from the formal development given in section 2. An investigation along these lines could also be of great interest for the study of measurement processes in the relativistic domain [23].

References

1. Feynman R. P., Vernon F. L. (1963): The Theory of a General Quantum System Interacting with a Linear Dissipative System. Ann. Phys. (N.Y.) **24**, 118-173.
2. Caldeira A. O., Leggett A. J. (1983): Quantum Tunneling in a Dissipative System. Ann. Phys. (N.Y.) **149**, 374-456; (1984) **153**, 445(E).
3. Gardiner C. W., Zoller P. (2000): *Quantum Noise.* 2nd. ed., Springer-Verlag, Berlin.
4. Zurek W. H. (1991): Decoherence and the transition from quantum to classical. Phys. Today **44**, 36-44.
5. Giulini D. *et al.* (1996): *Decoherence and the Appearence of a Classical World in Quantum Theory.* Springer-Verlag, Berlin.
6. Brune M. *et al.* (1996): Observing the progressive decoherence of the "meter" in a quantum measurement. Phys. Rev. Lett. **77**, 4887-4890.
7. Myatt C. J. *et al.* (2000): Decoherence of quantum superpositions through coupling to engineered reservoirs. Nature **403**, 269-273.
8. Weinberg S. (1996): *The Quantum Theory of Fields*, Volume I, Foundations. Cambridge University Press, Cambridge.
9. Jauch J. M., Rohrlich F. (1980): *The Theory of Photons and Electrons.* Springer-Verlag, New York.
10. Cohen–Tannoudji C., Dupont–Roc J., Grynberg G. (1998): *Atom-Photon Interactions.* John Wiley, New York.
11. Chou K.-c., Su Z.-b., Hao B.-l., Yu, L. (1985): Equilibrium and Nonequilibrium Formalisms Made Unified. Phys. Rep. **118**, 1-131.
12. Diósi L. (1990): Landau's Density Matrix in Quantum Electrodynamics. Found. Phys. **20**, 63-70.
13. Feynman R. P., Hibbs A. R. (1965): *Quantum Mechanics and Path Integrals.* McGraw-Hill, New York.
14. Grabert H., Schramm P., Ingold G.-L. (1988): Quantum Brownian Motion: The Functional Integral Approach. Phys. Rep. **168**, 115-207.
15. Barone P. M. V. B., Caldeira A. O. (1991): Quantum mechanics of radiation damping. Phys. Rev. **A43**, 57-63.
16. Anglin J. R., Paz J. P., Zurek W. H. (1997): Deconstructing Decoherence. Phys. Rev. **A55**, 4041-4053.

17. Jackson J. D. (1999): *Classical Electrodynamics.* Third Edition, John Wiley, New York.
18. Gradshteyn I. S., Ryzhik I. M. (1980): *Table of Integral, Series, and Products.* Academic Press, New York.
19. Dürr D., Spohn H. (2000): Decoherence Trough Coupling to the Radiation Field. In: Blanchard Ph., Giulini D., Joos E., Kiefer C., Stamatescu I.-O. (Eds.) *Decoherence: Theoretical, Experimental, and Conceptual Problems.* Springer-Verlag, Berlin, 77-86.
20. Joos E., Zeh H. D. (1985): The Emergence of Classical Properties Through Interaction with the Environment. Z. Phys. **B59**, 223-243.
21. Ghirardi G. C., Pearle P., Rimini A. (1990) Markov processes in Hilbert space and continuous spontaneous localization of systems of identical particles. Phys. Rev. **A42**, 78-89.
22. Ghirardi G. C., Grassi R., Pearle P. (1990): Relativistic Dynamical Reduction Models: General Framework and Examples. Found. Phys. **20**, 1271-1316.
23. Breuer H. P., Petruccione F. (1999): Stochastic Unravelings of Relativistic Quantum Measurements. In: Breuer H. P., Petruccione F. (Eds.) (1999): *Open Systems and Measurement in Relativistic Quantum Theory.* Springer-Verlag, Berlin, 81-116.

Decoherence: A Dynamical Approach to Superselection Rules?

Domenico Giulini

Theoretische Physik, Universität Zürich,
Winterthurerstrasse 190, CH-8057 Zürich, Switzerland

Abstract. It is well known that the dynamical mechanism of decoherence may cause apparent superselection rules, like that of molecular chirality. These 'environment-induced' or 'soft' superselection rules may be contrasted with 'hard' superselection rules, like that of electric charge, whose existence is usually rigorously demonstrated by means of certain symmetry principles. We address the question of whether this distinction between 'hard' and 'soft' is well founded and argue that, despite first appearance, it might not be. For this we first review in detail some of the basic structural properties of the spaces of states and observables in order to establish a fairly precise notion of superselection rules. We then discuss two examples: 1.) the Bargmann superselection rule for overall mass in ordinary quantum mechanics, and 2.) the superselection rule for charge in quantum electrodynamics.

1 Introduction

To explain the (apparent) absence of interferences between macroscopically interpretable states – like states describing spatially localized objects – is the central task for any attempt to resolve the measurement problem. First attempts in this direction just imposed additional rules, like that of the Copenhagen school, who *defined* a measurement device as a system whose state-space is classical, in the sense that the superposition principle is fully broken: superpositions between any two states simply do not exist. In a more modern language this may be expressed by saying that any two states of such a system are *disjoint*, i.e., separated by a superselection rule (see below). Proper quantum mechanical systems, which in isolation do obey the superposition principle, can then inherit superselection rules when coupled to such classical measurement devices.

Whereas there can be no doubt that the notion of classicality, as we understand it here, is mathematically appropriately encoded in the notions of *disjointness* and *superselection rules*, there still remains the physical question how these structures come to be imposed. In particular, if one believes that fundamentally all matter is described by some quantum theory, there is no room for an independent classical world. Classicality should be a feature that is emerging in accordance with, and not in violation of, the basic rules of quantum mechanics. This is the initial credo of those who believe in the program of *decoherence* [16], which aims to explain classicality by means of taking into account dynamical interactions with ambient systems, like the ubiquitous natural environment of the situation in question. (Note: It is not claimed to resolve the full measurement problem.) This leads to the notion of 'environment-induced superselection rules' [40].

Fundamental to the concept of dynamical decoherence is the notion of 'delocalization' [24]. The intuitive idea behind this is that through some dynamical process certain state characteristics ('phase relations'), which were locally accessible at one time, cease to be locally accessible in the course of the dynamical evolution. Hence locally certain superpositions cannot be verified anymore and an apparent obstruction to the superposition principle results. Such mechanisms are considered responsible for the above mentioned environment-induced superselection rule, of which a famous physical example is that of molecular chirality (see e.g. [38] and references therein). It has been established in many calculations of realistic situations that such dynamical processes of delocalization can be extremely effective over short time scales. But it is also intuitively clear that, mathematically speaking, it will never be strict in any finite time. Hence one will have to deal with notions of approximate- respectively asymptotic (for $t \to \infty$) superselection rules and disjointness of states [31,26], which needs some mathematical care.

Since for finite times such dynamical superselection rules are only approximately valid, they are sometimes called 'soft'. In contrast, 'hard' superselection rules are those which are rigorously established mathematical results within the kinematical framework of the theory, usually based on symmetry principles (see section 3 below), or on first principles of local QFT, like in the proof for the superselection rule for electric charge [33]. Such presentations seem to suggest that there is no room left for a dynamical origin of 'hard' superselection rules, and that hence these two notions of superselection rules are really distinct. However, we wish to argue that at least *some* of the existing proofs for 'hard' superselection rules give a false impression, and that quite to the contrary they actually need some dynamical input in order to be physically convincing. We will look at the case of Bargmann's superselection rule for total mass in ordinary quantum mechanics (which is clearly more of an academic example) and that of charge in QED. The discussion of the latter will be heuristic insofar as we will pretend that QED is nothing but quantum mechanics (in the Schrödinger representation) of the infinite-dimensional (constrained) Hamiltonian system given by classical electrodynamics. For a brief but general orientation on the subject of superselection rules and the relevant references we refer to Wightman's survey [39].

Let us stress again that crucial to the ideas presented here is of course that 'delocalized' does not at all mean 'destroyed', and that hence the loss of quantum coherence is only an *apparent* one. This distinction might be considered irrelevant FAPP (for all practical purposes) but it is important in attempts to understand apparent losses of quantum coherence *within* the standard dynamical framework of quantum mechanics.

As used here, the term 'local' usually refers to locality in the (classical) configuration space Q of the system, where we think of quantum states in the Schrödinger representation, i.e., as L^2-functions on Q. Every parametrization of Q then defines a partition into 'degrees of freedom'. Locality in Q is a more general concept than locality in ordinary physical space, although the latter forms a particular and physically important special case. Moreover, on a slightly more abstract level, one realizes that the most general description of why decoherence appears to occur is that only a *restricted* set of so-called physical observables are at ones disposal, and that *with respect to those* the relevant 'phase relations' *apparently* fade out of existence. It is sometimes convenient to express this by saying that decoherence occurs only with respect (or relative) to a 'choice' of observables [27]. Clearly this 'choice' is not meant to be completely free, since it has to be compatible with the dynamical laws and the physically realizable couplings (compare [24]). (In this respect the situation bears certain similarities to that of 'relevant' and 'irrelevant' degrees of freedom in statistical mechanics.) But to fully control those is a formidable task – to put it mildly. In any case it will be necessary to assume some a priori characterizations of what mathematical objects correspond to observables, and to do this in such a general fashion that one can effectively include superselection rules. This will be done in the next section.

2 Elementary Concepts

In this section we wish to convey a feeling for some of the concepts underlying the notion of superselection rules. We will take some care and time to do this, since many misconceptions can (and do!) arise from careless uses of these concepts. To gain intuition it is sometimes useful to dispense with some technicalities associated with infinite dimensions and continuous spectra and just look at finite dimensional situations; we will follow this strategy where indicated. We use the following, generally valid notations: $\mathcal{H}$ denotes a Hilbert space, $B(\mathcal{H})$ the algebra of bounded operators on $\mathcal{H}$. The antilinear operation of taking the hermitean conjugate is denoted by $*$ (rather than $\dagger$), which makes $B(\mathcal{H})$ a $*$-algebra. Given a set $\{A_\lambda\}$ where $\lambda \in \Lambda$ (= some index set), then by $\{A_\lambda\}'$ we denote the 'commutant' of $\{A_\lambda\}$ in $B(\mathcal{H})$, defined by

$$\{A_\lambda\}' := \{B \in B(\mathcal{H}) \mid BA_\lambda = A_\lambda B,\ \forall \lambda \in \Lambda\}. \tag{1}$$

Note that if the set $\{A_\lambda\}$ is left invariant under the $*$-map (in this case we call the set 'self-adjoint'), then $\{A_\lambda\}'$ is a $*$-subalgebra of $B(\mathcal{H})$. Also, the definition (1) immediately implies that

$$\mathcal{A} \subseteq \mathcal{B} \Rightarrow \mathcal{B}' \subseteq \mathcal{A}'. \tag{2}$$

2.1 Superselection Rules

There are many different ways to give a structural definition of superselection rules. Some stress the notion of *state* others the notion of *observable*. Often this dichotomy seems to result in very different attitudes towards the fundamental significance of superselection rules. This really seems artificial in a quantum mechanical context. In quantum *field* theory, i.e., if the underlying classical system has infinitely many degrees of freedom, the situation seems more asymmetric. This is partly due to the mathematical difficulties to define the full analog of the Schrödinger representation, i.e., to just construct the Hilbert space of states as L^2 space over the classical configuration space. In this paper we will partly ignore this mathematical difficulty and proceed heuristically by assuming that such a Schrödinger representation (of QED) exists to some level of rigour.

In traditional quantum mechanics, which stresses the notion of state, a system is fundamentally characterized by a Hilbert space, $\mathcal{H}$, the vectors of which represent (pure) states. We say 'represent' because this labeling by states through vectors is redundant: non-zero vectors which differ by an overall complex number label the same state, so that states can also be labeled by rays. We will use $\mathcal{PH}$ to denote the space of rays in $\mathcal{H}$. In many cases of interest this Hilbert space is of course just identified with the space of L^2-functions over the classical configuration space. Now, following the original definition given by W^3 [35], we say that a superselection rule operates on $\mathcal{H}$, if not all rays represent pure states, but only those which lie entirely in certain mutually orthogonal subspaces $\mathcal{H}_i \subset \mathcal{H}$, where

$$\mathcal{H} = \bigoplus_i \mathcal{H}_i. \tag{3}$$

The only rays which correspond to pure states are those in the disjoint union

$$\bigcup_i \mathcal{PH}_i\,. \tag{4}$$

Since no vector which lies skew to the partition (3) can, by assumption, represent a pure state, the superposition principle must be restricted to the $\mathcal{H}_i$. Moreover, since observables map pure states to pure states, they must leave the $\mathcal{H}_i$ invariant and hence all matrix-elements of observables between vectors from different sectors vanish. The $\mathcal{H}_i$ are called *coherent sectors* if the observables act irreducibly on them, i.e., if no further decomposition is possible; this is usually implied if a decomposition (3) is written down. States which lie in different coherent sectors are called *disjoint.* Note that disjointness of states is essentially also a statement about observables, since it means orthogonality of the original states *and* the respective states created from those with *all* observables. The existence of disjoint states is the characteristic feature of superselection rules.

From this we see that a partition (3) into coherent sectors implies that the set of physical observables is strictly smaller than the set of all self-adjoint (w.l.o.g. bounded) operators on $\mathcal{H}$. It can be characterized by saying that observables are those self adjoint operators on $\mathcal{H}$ which commute with the orthogonal projectors $P_i : \mathcal{H} \to \mathcal{H}_i$. So the P_i are themselves observables and generate the center (see below) of the algebra of observables.

This suggests a 'dual', more algebraic way to look at superselection rules, which starts with the algebra of observables $\mathcal{O}$. Then superselection rules are said to occur if the algebra of observables, $\mathcal{O}$, – which we think of as being given by bounded operators on some Hilbert space $\mathcal{H}$[1] – has a non-trivial center $\mathcal{O}^c$. Recall that

$$\mathcal{O}^c := \{A \in \mathcal{O} \mid AB = BA, \ \forall B \in \mathcal{O}\}. \tag{5}$$

Suppose $\mathcal{O}^c$ is generated by self-adjoint elements $\{C_\mu, \mu = 1, 2, ..\}$ which have simultaneous eigenspaces $\mathcal{H}_i$, then the $\mathcal{H}_i$'s are just the coherent sectors. Indeed, as already remarked, matrix elements of operators from $\mathcal{O}$ between states from different coherent sectors (i.e. differing in the eigenvalue of at least one C_μ) necessarily vanish. Thus if ϕ_1 and ϕ_2 are two non-zero vectors from $\mathcal{H}_i$ and $\mathcal{H}_j$ with $i \neq j$, their superposition $\phi := \phi_i + \phi_2$ defines a state whose density matrix $\rho := P_\phi$ (=orthogonal projector onto the ray generated by ϕ) satisfies

$$\mathrm{tr}(\rho A) = \mathrm{tr}((\lambda_1 \rho_1 + \lambda_2 \rho_2) A), \quad \forall A \in \mathcal{O}, \tag{6}$$

where $\lambda_{1,2} = \|\phi_{1,2}\|^2 / \|\phi\|^2$ and $\rho_{1,2} = P_{\phi_{1,2}}$. This means that ρ is a non-pure state of $\mathcal{O}$, since it can be written as a non-trivial convex combination of other density matrices. Hence we come back to the statements expressed by (3) and (4). Also note the following: in quantum mechanics the decomposition of a non-pure density matrix as a convex combination of pure density matrices – the so-called extremal decomposition – is generically not unique, thus preventing the (ignorance-) interpretation as statistical "mixtures".[2] However, for the special density matrices of the form $\rho = |\phi\rangle\langle\phi|$, where $|\phi\rangle \in \mathcal{H}$, the

[1] In Algebraic Quantum Mechanics one associates to each quantum system an abstract C^*-algebra, $\mathcal{C}$, which is thought of as being the mathematical object that fully characterizes the system in isolation, i.e. its intrinsic or 'ontic' properties. But this is not yet what we call the algebra of observables. This latter algebra is not uniquely determined by the former. It is obtained by studying faithful representations of $\mathcal{C}$ in some Hilbert space $\mathcal{H}$, such that $\mathcal{C}$ can be identified with some subalgebra of $B(\mathcal{H})$ (the bounded operators on $\mathcal{H}$). This is usually done by choosing a reference state (positive linear functional) on $\mathcal{C}$ and performing the GNS construction. Then $\mathcal{C}$ inherits a norm which is used to close $\mathcal{C}$ (as topological space) in $B(\mathcal{H})$. It is this resulting algebra which corresponds to our $\mathcal{O}$. Technically speaking it is a von Neumann algebra which properly contains an embedded copy of $\mathcal{C}$. The added observables (those in $\mathcal{O} - \mathcal{C}$) do not describe intrinsic but *contextual* properties. For example, it may happen that $\mathcal{O}$ has non-trivial center whereas $\mathcal{C}$ doesn't. In this case the superselection rules described by $\mathcal{O}$ are contextual. See [31] for a more extended discussion of this point.

[2] Hence the term 'mixture' for a non-pure state is misleading since we cannot tell the components and hence have no ensemble interpretation. For this reason we will say 'non-pure state' rather than 'mixture'.

extremal decomposition *is* unique and given by $\phi = \sum_i \lambda_i P_{\phi_i}$, where ϕ_i is the orthogonal projection of ϕ into $\mathcal{H}_i$, P_{ϕ_i} the orthogonal projector onto ϕ_i's ray, and $\lambda_i = \|\phi_i\|^2/\|\phi\|^2$. This is the relevance of superselection rules for the measurement problem: to produce *unique* extremal decompositions – and hence statistical 'mixtures' in the proper sense of the word – into an ensemble of pure states. There is a long list of papers dealing with the mathematical problem of how superselection sectors can arise dynamically; see e.g. [19,30,2,26] and the more general discussions in [27,31].

2.2 Dirac's Requirement

Dirac was the first who spelled out certain rules concerning the spaces of states and observables [6]. He defined the notion of *compatible* (i.e., simultaneously performable) *observations*, which mathematically are represented by a set of commuting observables, and the notion of a *complete set* of such observables, which is meant to say that there is precisely one state for each set of simultaneous "eigenvalues". Starting from the hypothesis that states are faithfully represented by rays, Dirac deduced that a complete set of such mutually compatible observables existed. But this only makes sense if all the observables in question have purely discrete spectra.

In the general case one has to proceed differently: We heuristically define Dirac's requirement as the statement, that *there exists at least one complete set of mutually compatible observables* and show how it can be rephrased mathematically so that it applies to all cases. In doing this we essentially follow Jauch's exposition [22]. To develop a feeling for what is involved, we will first describe some of the consequences of Dirac's requirement in the most simplest case: a finite dimensional Hilbert space. We will use this insight to rephrase it in such a way to stay generally valid in infinite dimensions.

Gaining intuition in finite dimensions. So let $\mathcal{H}$ be an n-dimensional complex Hilbert-space, then $B(\mathcal{H})$ is the algebra of complex $n \times n$ matrices. Physical observables are represented by hermitean matrices in $B(\mathcal{H})$, but we will explicitly *not* assume the converse, namely that *all* hermitean matrices correspond to physical observables. Rather we assume that the physical observables are somehow given to us by some set $\mathcal{S}$ of hermitean matrices. This set does not form an algebra, since taking products and complex linear combinations does not preserve hermiticity. But for mathematical reasons it would be convenient to have such an algebraic structure, and just work with the algebra $\mathcal{O}$ generated by this set, called the *algebra of observables.* [Note the usual abuse of language, since only the hermitean elements in $\mathcal{O}$ are observables.] But for this replacement of $\mathcal{S}$ by $\mathcal{O}$ to be allowed $\mathcal{S}$ must have been a set of hermitean matrices which is uniquely determined by $\mathcal{O}$, for otherwise we can not reconstruct the set $\mathcal{S}$ from $\mathcal{O}$. To grant us this mathematical convenience we assume that $\mathcal{S}$ was already maximal, i.e. that $\mathcal{S}$ already contains all the hermitean matrices that it generates. But we stress that there seems to be no obvious reason why in a particular practical situation the set of physically realizable observables should be maximal in this sense.

We may choose a set $\{O_1, \ldots O_m\}$ of hermitean generators of $\mathcal{O}$. Then $\mathcal{O}$ may be thought of as the set of all complex polynomials in these (generally non-commuting) matrices. But note that we need not consider higher powers than $(n-1)$ of each O_i, since each complex $n \times n$ matrix O is a zero of its own characteristic polynomial p_O, i.e. satisfies $p_O(O) = 0$, by the theorem of Cayley-Hamilton. Since this polynomial is of order n, O^n can be re-expressed by a polynomial in O of order at most $(n-1)$. For example, the $*$-algebra generated by a single hermitean matrix O can be identified with the set of all polynomials of degree at most $(n-1)$ and whose multiplication law is as usual, followed by the procedure of reducing all powers n and higher of O via $p_O(O) = 0$.

Now let $\{A_1, \cdots, A_m\} =: \{A_i\}$ be a complete set of mutually commuting observables. It is not difficult to show that there exists an observable A and polynomials p_i, $i = 1, \cdots, m$ such that $A_i = p_i(A)$ (see [20] for a simple proof). This actually means that the algebra generated by $\{A_i\}$ is just the n-dimensional algebra of polynomials of degree at most $n-1$ in A (see below for justification), which we call $\mathcal{A}$. This algebra is abelian, which is equivalently expressed by saying that $\mathcal{A}$ is contained in its commutant (compare (1)):

$$\mathcal{A} \subseteq \mathcal{A}' \qquad \boxed{\text{`}\mathcal{A}\text{ is abelian'}} \tag{7}$$

Now comes the requirement of completeness. In terms of A it is easy to see that it is equivalent to the condition that A has a simple spectrum (i.e. the eigenvalues are pairwise distinct). This has the following consequence: Let B be an observable that commutes with A, then B is also a function of A, i.e., $p_B(B) = A$ for some polynomial p_B. The proof is simple: We simultaneously diagonalize A and B with eigenvalues α_a and β_a, $a = 1, \cdots, n$. We wish to find a polynomial of degree $n-1$ such that $p_B(\alpha_a) = \beta_a$. Writing $p_B(x) = a_{n-1}x^{n-1} + \cdots + a_0$, this leads to a system of n linear equations ($\alpha_a^b :=$ b^{th} power of α_a)

$$\sum_{b=0}^{n-1} \alpha_a^b a_b = \beta_a, \qquad \text{for } a = 1, \ldots, n, \tag{8}$$

for the n unknowns $(a_0, \cdots, a_{n-1})$. Its determinant is of course just the Vandermonde determinant for the n tuple $(\alpha_1, \cdots, \alpha_n)$:

$$\det\{\alpha_a^b\} = \prod_{a<b} (\alpha_a - \alpha_b), \tag{9}$$

which is non-zero if and only if (=iff) A's spectrum is simple. This implies that every observable that commutes with $\mathcal{A}$ is already contained in $\mathcal{A}$. (It follows from this that the algebra generated by $\{A_i\}$ is equal to, and not just a subalgebra of, the algebra generated by $\{A\}$, as stated above.) Since a $*$-algebra is generated by its self-adjoint elements (observables), $\mathcal{A}$ cannot be properly enlarged as abelian $*$-algebra by adding more commuting generators. In other words, $\mathcal{A}$ is *maximal*. Since $\mathcal{A}'$ is a $*$-algebra, this can be equivalently expressed by

$$\mathcal{A}' \subseteq \mathcal{A} \qquad \boxed{\text{`}\mathcal{A}\text{ is maximal'}} \tag{10}$$

Equations (7) and (10) together are equivalent to Dirac's condition, which can now be stated in the following form, first given by Jauch [22]: the algebra of observables $\mathcal{O}$ contains a maximal abelian $*$-subalgebra $\mathcal{A} \subseteq \mathcal{O}$, i.e.,

$$\boxed{\text{Dirac's requirement, 1}^{\text{st}}\text{ version: } \exists\ \mathcal{A} \subseteq \mathcal{O} \text{ satisfying } \mathcal{A} = \mathcal{A}'} \tag{11}$$

This may seem as if Dirac's requirement could be expressed in purely algebraic terms. But this is deceptive, since the very notion of 'commutant' (compare (1)) makes reference to the Hilbert space $\mathcal{H}$ through $B(\mathcal{H})$. Without further qualification the term 'maximal' always means maximal *in* $B(\mathcal{H})$.[3]

This reference to $\mathcal{H}$ can be further clarified by yet another equivalent statement of Dirac's requirement. Since $\mathcal{A}$ consists of polynomials in the observable A, which has a simple spectrum, the following is true: there exists a vector $|g\rangle \in \mathcal{H}$, such that for *any* vector $\phi \in \mathcal{H}$ there exists a polynomial p_ϕ such that

$$p_\phi(A)|g\rangle = |\phi\rangle. \tag{12}$$

Such a vector $|g\rangle$ is called a generating or *cyclic vector* for $\mathcal{A}$ in $\mathcal{H}$. The proof is again very simple: let $\{\phi_1, \cdots, \phi_n\}$ be the pairwise distinct, non-zero eigenvectors of A (with any normalization); then choose

$$|g\rangle = \sum_{i=1}^{n} |\phi_i\rangle. \tag{13}$$

Equation (12) now defines again a system of n linear equations for the n coefficients $a_{n-1}, \cdots, a_0$ of the polynomial p_ϕ, whose determinant is again the Vandermonde determinant (9) for the n eigenvalues $\alpha_1, \cdots, \alpha_n$ of A. Conversely, if A had an eigenvalue, say α_1, with eigenspace $\mathcal{H}_1$ of two or higher dimensions, then such a cyclic $|g\rangle$ cannot exist. To see this, suppose it did, and let $|\phi_1^\perp\rangle \in \mathcal{H}_1$ be orthogonal to the projection of $|g\rangle$ into $\mathcal{H}_1$. Then $\langle\phi_1^\perp|p(A)g\rangle = 0$ for all polynomials p. Thus $|\phi_1^\perp\rangle$ is unreachable, contradicting our initial assumption. Hence a simple spectrum of A is equivalent to the existence of a cyclic vector.

[3] The condition for an abelian $\mathcal{A} \subseteq \mathcal{O}$ to be maximal in $\mathcal{O}$ would be $\mathcal{A} = \mathcal{A}' \cap \mathcal{O}$. Such abelian subalgebras *always* exist, in contrast to those $\mathcal{A} \subseteq \mathcal{O}$ which satisfy the stronger condition to be maximal in the ambient algebra $B(\mathcal{H})$.

The general case. In infinite dimensions we have to care a little more about the topology on the space of observables, since here there are many inequivalent ways to generalize the finite dimensional case. The natural choice is the so-called 'weak topology', which is characterized by declaring that a sequence $\{A_i\}$ of observables converges to the observable A if the sequence $\langle\phi|A_i|\psi\rangle$ of complex numbers converges to $\langle\phi|\mathcal{A}|\psi\rangle$ for all $|\phi\rangle, |\psi\rangle \in \mathcal{H}$. Hence one also requires that the algebra of observables is weakly closed (i.e., closed in the weak topology). Such a weakly closed $*$-subalgebra of $B(\mathcal{H})$ is called a W^*- or von-Neumann-algebra (we shall use the first name for brevity).

A crucial and extremely convenient point is, that the weak topology is fully encoded in the operation of taking the commutant (see (1)), in the following sense: Let $\{A_\lambda\}$ be any subset of $B(\mathcal{H})$, then $\{A_\lambda\}'$ is automatically weakly closed (see [22] p 716 for a simple proof) and hence a W^*-algebra. Moreover, the weak closure of a $*$-algebra $\mathcal{A} \subseteq B(\mathcal{H})$ is just given by $\mathcal{A}''$ (the commutant of the commutant). Hence we can characterize a W^*-algebra purely in terms of commutants: $\mathcal{A}$ is W^* iff $\mathcal{A} = \mathcal{A}''$.

This allows to easily generalize the notion of 'algebra generated by observables': Let $\{O_\lambda\}$ be a set of self-adjoint elements in $B(\mathcal{H})$, then $\mathcal{O} := \{O_\lambda\}''$ is called the (W^*-) algebra generated by this set. This definition is natural since $\{O_\lambda\}''$ is easily seen to be the smallest W^*-algebra containing $\{O_\lambda\}$, for if $\{O_\lambda\} \subseteq \mathcal{B} \subseteq \mathcal{O}$ for some W^*-algebra $\mathcal{B}$, then taking the commutant twice yields $\mathcal{B} = \mathcal{O}$.[4]

Now we see that Dirac's requirement in the form (11) directly translates to the general case if all algebras involved (i.e. $\mathcal{A}$ and $\mathcal{O}$) are understood as W^*-algebras. Now we also know what a 'complete set of (bounded) commuting observables' is, namely a set $\{A_\lambda\} \subseteq B(\mathcal{H})$ whose generated W^*-algebra $\mathcal{A} := \{A_\lambda\}''$ is maximal abelian: $\mathcal{A} = \mathcal{A}'$. This latter condition is again equivalent to the existence of a cyclic vector $|g\rangle \in \mathcal{H}$ for $\mathcal{A}$, where in infinite dimensions the definition of cyclic is that $\{\mathcal{A}|g\rangle\}$ is *dense* in (rather than equal to) $\mathcal{H}$. It is also still true that there is an observable A such that all A_λ are functions (in an appropriate sense, not just polynomials of course) of A [34]. But since A's spectrum may be (partially) continuous, there is no direct interpretation of a 'simple' spectrum as in finite dimensions. Rather, one now defines simplicity of the spectrum of A by the existence of a cyclic vector for $\mathcal{A} = \{A\}''$.

Now we come to our final reformulation of Dirac's condition. Namely, looking at (11), we may ask whether we could not reformulate the existence of such a maximal abelian $\mathcal{A}$ purely in terms of the algebra of observables $\mathcal{O}$ alone. This is indeed possible. We have $\mathcal{A} \subseteq \mathcal{O} \Rightarrow \mathcal{O}' \subseteq \mathcal{A}' = \mathcal{A} \subseteq \mathcal{O}$, hence $\mathcal{O}' \subseteq \mathcal{O}$. Since $\mathcal{O} = \mathcal{O}''$ the last condition is equivalent to saying that $\mathcal{O}'$ is abelian ($\mathcal{O}' \subseteq \mathcal{O}''$), or to saying that $\mathcal{O}'$ is the center $\mathcal{O}^c$ of $\mathcal{O}$, since by (1) and

[4] Note: for any $M \subseteq B(\mathcal{H})$ definition (1) immediately yields $M \subseteq M''$ and hence $M' \supseteq M'''$ (by (2)). But also $M' \subseteq M'''$ (by replacing $M \rightarrow M'$); therefore $M' = M'''$ for any $M \subseteq B(\mathcal{H})$.

(5) the center can be written as $\mathcal{O}^c = \mathcal{O} \cap \mathcal{O}'$. Now, conversely, it was shown in [23] that an abelian $\mathcal{O}'$ implies the existence of a maximal abelian $\mathcal{A} \subseteq \mathcal{O}$. Hence we have the following alternative formulation of Dirac's requirement, first spelled out, independently of (11), by Wightman [37], who called it the 'hypothesis of commutative superselection rules':

$$\boxed{\text{Dirac's requirement, 2}^{\text{nd}}\text{ version: } \mathcal{O}' \text{ is abelian}} \tag{14}$$

There are several interesting ways to interpret this condition. From its derivation we know that it is equivalent to the existence of a maximal abelian $\mathcal{A} \subseteq \mathcal{O}$. But we can in fact make an apparently stronger statement, which also relates to the earlier footnote 3, namely: (14) is equivalent to the condition, that *any* abelian $\mathcal{A} \subseteq \mathcal{O}$ that is maximal in $\mathcal{O}$, i.e. satisfies $\mathcal{A} = \mathcal{A}' \cap \mathcal{O}$, is also maximal in $B(\mathcal{H})$.[5]

2.3 Dirac's Condition and Gauge Symmetries

Another way to understand (14) is via its limitations on *gauge-symmetries*. To see this, we mention that any W^*-algebra is generated by its unitary elements. Hence $\mathcal{O}'$ is generated by a set $\{U_\lambda\}$ of unitary operators. Each U_λ commutes with *all* observables and therefore generates a one-parameter group of gauge-transformations. Condition (14) is then equivalent to saying that the total gauge group, which is generated by all U_λ, is *abelian*. Note also that an abelian $\mathcal{O}'$ implies that the gauge-algebra, $\{U_\lambda\}'' = \mathcal{O}'$, is contained in the observables, $\mathcal{O}' \subseteq \mathcal{O}'' = \mathcal{O}$, so that $\mathcal{O}' = \mathcal{O}^c$. From this one can infer the following central statement:

$$\boxed{\begin{array}{l}\text{Dirac's requirement implies that gauge- and}\\ \text{sectorial structures are fully determined by the}\\ \text{center } \mathcal{O}^c \text{ of the algebra of observables } \mathcal{O}.\end{array}} \tag{15}$$

To see in what sense this is true we remark that for W^*-algebras we can simultaneously diagonalize all observables in $\mathcal{O}^c$. That means that we can write $\mathcal{H}$ in an essentially unique way as direct integral over the real line of Hilbert spaces $\mathcal{H}(\lambda)$ using some (Lebesgue-Stieltjes-) measure σ:

$$\mathcal{H} = \int_{\mathbb{R}}^{\oplus} d\sigma(\lambda)\, \mathcal{H}(\lambda). \tag{16}$$

Operators in $\mathcal{O}$ respect this decomposition in the sense that each $O \in \mathcal{O}$ acts on $\mathcal{H}$ componentwise via some bounded operator $O(\lambda)$ on $\mathcal{H}(\lambda)$. If $O \in \mathcal{O}^c$

[5] Proof: We need to show that ($\mathcal{O}'$ abelian) $\Leftrightarrow$ ($\mathcal{A} = \mathcal{A}' \cap \mathcal{O} \Rightarrow \mathcal{A} = \mathcal{A}'$). '$\Rightarrow$': $\mathcal{O}'$ abelian implies $\mathcal{O}' \subseteq \mathcal{O}'' = \mathcal{O}$ and $\mathcal{A} \subseteq \mathcal{O}$ implies $\mathcal{O}' \subseteq \mathcal{A}'$, so that $\mathcal{O}' \subseteq \mathcal{A}' \cap \mathcal{O}$. Hence $\mathcal{A} = \mathcal{A}' \cap \mathcal{O}$ implies $\mathcal{O}' \subseteq \mathcal{A}$, which implies $\mathcal{A}' \subseteq \mathcal{O}'' = \mathcal{O}$, and hence $\mathcal{A} = \mathcal{A}'$. '$\Leftarrow$': ($\mathcal{A} = \mathcal{A}' \cap \mathcal{O} \Rightarrow \mathcal{A} = \mathcal{A}'$) is equivalent to $\mathcal{A}' \subseteq \mathcal{O}$, which implies $\mathcal{O}' \subseteq \mathcal{A}'' = \mathcal{A}$ and hence that $\mathcal{O}'$ is abelian.

then each $O(\lambda)$ is a multiple $\phi(\lambda) \in \mathbb{C}$ of the unit operator. Moreover, the set of all $\{\mathcal{O}(\lambda)\}$ induced from $\mathcal{O}$ for each fixed λ acts irreducibly on $\mathcal{H}(\lambda)$.[6] Hence, provided that Dirac's requirement is satisfied, (16) is the generally valid version of (3). The notion of disjointness now acquires an intuitive meaning: two states $|\Psi_1\rangle$ and $|\Psi_2\rangle$ are separated by a superselection rule (are disjoint), iff their component-state-functions $\lambda \to |\psi_1(\lambda)\rangle$ and $\lambda \to |\psi_2(\lambda)\rangle$ have disjoint support on $\mathbb{R}$ (up to measure-zero sets). Note that by spectral decomposition the superselection observables can be decomposed into the projectors in $\mathcal{O}^c$, which for (16) are all given by multiplications with characteristic functions $\chi(\lambda)$ for σ-measurable sets in $\mathbb{R}$.

Non-abelian gauge groups We have seen that the fulfillment of Dirac's requirement allows to give a full structural characterisation for the spaces of (pure) states and observables. How general is this result? Does it exclude cases of physical interest? At first glance this seems indeed to be the case: just consider a situations with non-abelian gauge groups; for example, the quantum mechanical system of $n > 2$ identical spinless particles with n-particle Hilbert space $\mathcal{H} = L^2(\mathbb{R}^{3n})$ on which the permutation group $G = S_n$ of n objects acts in the obvious way by unitary operators $U(g)$. That these particles are identical means that observables must commute with each $U(g)$. Without further restrictions on observables one would thus define $\mathcal{O} := \{U(g),\, g \in G\}'$. Hence $\mathcal{O}'$ is the W^*-algebra generated by all $U(g)$, which is clearly non-abelian, thus violating (14). But does this generally imply that general particle statistics cannot be described in a quantum-mechanical setting which fulfills Dirac's requirement? The answer to this question is 'no'. Let us explain why.

If we decompose $\mathcal{H}$ according to the unitary, irreducible representations of G we obtain ([10,14])

$$\mathcal{H} = \bigoplus_{i=1}^{p(n)} \mathcal{H}_i\,, \tag{17}$$

where i labels the $p(n)$ inequivalent, unitary, irreducible representations D_i of G of dimension d_i. Each $\mathcal{H}_i$ has the structure $\mathcal{H}_i \cong \mathbb{C}^{d_i} \otimes \tilde{\mathcal{H}}_i$, where G acts irreducibly via D_i on $\mathbb{C}^{d_i}$ and trivially on $\tilde{\mathcal{H}}_i$ whereas $\mathcal{O}$ acts irreducibly via some $*$-representation π_i on $\tilde{\mathcal{H}}_i$ and trivially on $\mathbb{C}^{d_i}$. π_i and π_j are inequivalent if $i \neq j$. Hence we see that $\mathcal{H}_i$ furnishes an irreducible representation for $\mathcal{O}$, iff $d_i = 1$, i.e., for the Bose and Fermi sectors only. Pure states from these sectors are just the rays in the corresponding $\mathcal{H}_i$. In contrast, for $d_i > 1$, given a non-zero vector $|\phi\rangle \in \tilde{\mathcal{H}}_i$, all non-zero vectors in the d_i-dimensional subspace $\mathbb{C}^{d_i} \otimes |\phi\rangle \subset \mathcal{H}_i$ define the *same* pure state, i.e., the same expectation-value-functional on $\mathcal{O}$. Furthermore, a vector in $\mathcal{H}_i \cong \mathbb{C}^{d_i} \otimes \tilde{\mathcal{H}}_i$ which is not a pure

[6] It is this irreducibility statement which depends crucially on the fulfillment of Dirac's requirement. In general, the $O(\lambda)$'s will act irreducibly on $\mathcal{H}(\lambda)$ for each λ, iff $\mathcal{O}^c$ is maximal abelian *in* $\mathcal{O}'$, i.e., iff $\mathcal{O}^c = (\mathcal{O}^c)' \cap \mathcal{O}'$. But we already saw that (14) also implies $\mathcal{O}^c = \mathcal{O}'$ so that this is fulfilled.

tensor product defines a non-pure state, since the restriction of $O \in \mathcal{O}$ to $\mathcal{H}_i$ is of the form $\mathbf{1} \otimes \tilde{O}$, which means that a vector in $\mathcal{H}_i$ defines a state given by the reduced density matrix obtained by tracing over the left (i.e. $\mathbb{C}^{d_i}$) state space. From elementary quantum mechanics we know that the resulting state is pure, iff the vector in $\mathcal{H}_i$ was a pure tensor product (i.e. of rank one). Hence in those $\mathcal{H}_i$ where $d_i > 1$ not all vectors correspond to pure states, and those which do represent pure states in a redundant fashion by higher dimensional subspaces, sometimes called 'generalized rays' in the older literature on parastatistics [28].

However, the factors $\mathbb{C}^{d_i}$ are completely redundant as far as physical information is concerned, which is already fully encoded in the irreducible representations π_i of $\mathcal{O}$ on $\tilde{\mathcal{H}}_i$; no further physical information is contained in d_i-fold repetitions of π_i. Hence we can define a new, truncated Hilbert space

$$\tilde{\mathcal{H}} := \bigoplus_{i=1}^{p(n)} \tilde{\mathcal{H}}_i \,. \tag{18}$$

This procedure has also been called 'elimination of the generalized ray' in the older literature on parastatistics [18] – see also [14] for a more recent discussion of this point. Since every pure state in $\mathcal{H}$ is also contained in $\tilde{\mathcal{H}}$, just without repetition, these two sets are called 'phenomenological equivalent' in the literature on QFT (e.g. in chapter 6.1.C of [4]). The point is that pure states are now faithfully labelled by rays in the $\tilde{\mathcal{H}}_i$ and that $\mathcal{O}'$ – where the commutant is now taken in $B(\tilde{\mathcal{H}})$ rather than $B(\mathcal{H})$ – is generated by $\mathbf{1}$ and the $p(n)$ (commuting!) projectors into the $\tilde{\mathcal{H}}_i$'s. Hence Dirac's requirement is satisfied. But clearly the original gauge group has no action on $\tilde{\mathcal{H}}$ anymore, but there is also no physical reason why one should keep it.[7] It served to define $\mathcal{O}$, but then only its irreducible representations π_i are of interest. Only a residual action of the center of G still exists, but the gauge group generated by the projectors into the $\tilde{\mathcal{H}}_i$ consists in fact of the continuous group of $p(n)$ copies of $U(1)$, one global phase change for each sector. Its meaning is simply to induce the separation into the different sectors $(\tilde{\mathcal{H}}_i, \pi_i)$, and that in accordance with Dirac's requirement.

To sum up, we have seen that even if a theory is initially formulated via non-abelian gauge groups, we can give it a physically equivalent formulation that has at most a residual abelian gauge group left and hence obeys Dirac's requirement. Hence the 'obvious' counterexamples to Dirac's requirement turn out to be harmless. This is generally true in quantum mechanics,

[7] In [10] Dirac's requirement together with the requirement that the physical Hilbert space must carry an action of the gauge group has been used to "prove" the absence of parastatistics. In our opinion there seems to be no physical reason to accept the second requirement and hence the "proof"; compare [18] and [14].

but in quantum field theory there are genuine possibilities to violate Dirac's condition which we will ignore here.[8]

3 Superselection Rules via Symmetry Requirements

The requirement that a certain group must act on the set of all physical states is often the (kinematical) source of superselection rules. Here I wish to explain the structure of this argument.

Note first that in quantum mechanics we identify the states of a closed system with rays and not with vectors which represent them (in a redundant fashion). It is therefore not necessary to require that a symmetry group G acts on the Hilbert space $\mathcal{H}$, but rather it is sufficient that it acts on $\mathcal{PH}$, the space of rays, via so-called ray-representations. Mathematically this is a non-trivial relaxation since not every ray-representation of a symmetry group G (i.e. preserving the ray products) lifts to a unitary action of G on $\mathcal{H}$. What may go wrong is not that for a given $g \in G$ we cannot find a unitary (or anti-unitary) operator U_g on $\mathcal{H}$; that is assured by Wigner's theorem (see [3] for a proof). Rather, what may fail to be possible is that we can choose the U_g's in such a way that we have an *action*, i.e., that $\mathsf{U}_{g_1}\mathsf{U}_{g_2} = U_{g_1 g_2}$. As is well known, this is precisely what happens for the implementation of the Galilei group in ordinary quantum mechanics. Without the admission of ray representations we would not be able to say that ordinary quantum mechanics is Galilei invariant.

To be more precise, to have a ray-representation means that for each $g \in G$ there is a unitary[9] transformation U_g which, instead of the usual representation property, are only required to satisfy the weaker condition

$$\mathsf{U}_{g_1}\mathsf{U}_{g_2} = \exp(\mathrm{i}\xi(g_1, g_2))\,\mathsf{U}_{g_1 g_2} \tag{19}$$

for some function $\xi : G \times G \to \mathbb{R}$, called multiplier exponent, satisfying[10]

$$\xi(1, g) = \xi(g, 1) \qquad = 0, \tag{20}$$

$$\xi(g_1, g_2) - \xi(g_1, g_2 g_3) + \xi(g_1 g_2, g_3) - \xi(g_2, g_3) = 0. \tag{21}$$

The second of these conditions is a direct consequence of associativity:

$$\mathsf{U}_{g_1}(\mathsf{U}_{g_2}\mathsf{U}_{g_3}) = (\mathsf{U}_{g_1}\mathsf{U}_{g_2})\mathsf{U}_{g_3}.$$

[8] An abelian $\mathcal{O}'$ implies that $\mathcal{O}$ is a von Neumann algebra of type I (see [7], chapter 8) whereas truly infinite systems in QFT are often described by type III algebras.

[9] For simplicity we ignore anti-unitary transformations. They cannot arise if, for example, G is connected.

[10] The following conditions might seem a little too strong, since it would be sufficient to require the equalities in (20) and (21) only mod 2π; this also applies to (22). But for our application in section 4 it is more convenient to work with strict equalities, which in fact implies no loss of generality; compare [32].

Obviously these maps project to an action of G on $\mathcal{PH}$. Any other lift of this action on $\mathcal{PH}$ onto $\mathcal{H}$ is given by a redefinition $\mathsf{U}_g \to \mathsf{U}'_g := \exp(i\gamma(g))\mathsf{U}_g$, for some function $\gamma : G \to \mathbb{R}$ with $\gamma(1) = 0$, resulting in new multiplier exponents

$$\xi'(g_1, g_2) = \xi(g_1, g_2) + \gamma(g_1) - \gamma(g_1 g_2) + \gamma(g_2), \tag{22}$$

which again satisfy (20) and (21). The ray representations U and U' are then said to be equivalent, since the projected actions on $\mathcal{PH}$ are the same. We shall also say that two multiplier exponents ξ, ξ' are equivalent if they satisfy (22) for some γ.

We shall now see how the existence of inequivalent multiplier exponents, together with the requirement that the group should act on the space of physical states, may clash with the superposition principle and thus give rise to superselection rules. For this we start from two Hilbert spaces $\mathcal{H}'$ and $\mathcal{H}''$ and actions of a symmetry group G on $\mathcal{PH}'$ and $\mathcal{PH}''$, i.e., ray representations U' and U'' on $\mathcal{H}'$ and $\mathcal{H}''$ up to equivalences (22). We consider $\mathcal{H} = \mathcal{H}' \oplus \mathcal{H}''$ and ask under what conditions does there exist an action of G on $\mathcal{PH}$ which restricts to the given actions on the subsets $\mathcal{PH}'$ and $\mathcal{PH}''$. Equivalently: when is $\mathsf{U} = \mathsf{U}' \oplus \mathsf{U}''$ a ray representation of G on $\mathcal{H}$ for some choice of ray-representations U' and U'' within their equivalence class? To answer this question, we consider

$$\begin{aligned} \mathsf{U}_{g_1}\mathsf{U}_{g_2} &= (\mathsf{U}'_{g_1} \oplus \mathsf{U}''_{g_1})(\mathsf{U}'_{g_2} \oplus \mathsf{U}''_{g_2}) \\ &= \exp(\mathrm{i}\xi'(g_1, g_2))\mathsf{U}'_{g_1 g_2} \oplus \exp(\xi''(g_1, g_2))\mathsf{U}''_{g_1 g_2} \end{aligned} \tag{23}$$

and note that this can be written in the form (19), for some choice of ξ', ξ'' within their equivalence class, iff the phase factors can be made to coincide, that is, iff ξ' and ξ'' are equivalent. This shows that there exists a ray-representation on $\mathcal{H}$ which restricts to the given equivalence classes of given ray representations on $\mathcal{H}'$ and $\mathcal{H}'$, iff the multiplier exponents of the latter are equivalent. Hence, if the multiplier exponents ξ' and ξ'' are *not* equivalent, the action of G cannot be extended beyond the disjoint union $\mathcal{PH}' \cup \mathcal{PH}''$. Conversely, *if* we require that the space of physical states must support an action of G, then non-trivial superpositions of states in $\mathcal{H}'$ and $\mathcal{H}''$ must be excluded from the space of (pure) physical states.

This argument shows that if we insist of implementing G as symmetry group, superselection rules are sometimes unavoidable. A formal trick to avoid them would be not to require G, but a slightly larger group, $\bar{G}$, to act on the space of physical states. $\bar{G}$ is chosen to be the group whose elements we label by (θ, g), where $\theta \in \mathbb{R}$, and the multiplication law is

$$\bar{g}_1 \bar{g}_2 = (\theta_1, g_1)(\theta_2, g_2) = (\theta_1 + \theta_2 + \xi(g_1, g_2), g_1 g_2). \tag{24}$$

It is easy to check that the elements of the form $(\theta, 1)$ lie in the center of $\bar{G}$ and form a normal subgroup $\cong \mathbb{R}$ which we call Z. Hence $\bar{G}/Z = G$ but G

need not be a subgroup of $\bar{G}$. $\bar{G}$ is a central $\mathbb{R}$ extension[11] of G (see e.g. [32]). Now a ray-representation U of G on $\mathcal{H}$ defines a proper representation U of $\bar{G}$ on $\mathcal{H}$ by setting

$$U_{(\theta,g)} := \exp(\mathrm{i}\theta)\mathsf{U}_g. \tag{25}$$

Then $\bar{G}$ is properly represented on $\mathcal{H}'$ and $\mathcal{H}''$ and hence also on $\mathcal{H} = \mathcal{H}' \oplus \mathcal{H}''$. The above phenomenon is mirrored here by the fact that Z acts trivially on $\mathcal{PH}'$ and $\mathcal{PH}''$ but non-trivially on $\mathcal{PH}$, and the superselection structure comes about by requiring physical states to be fixed points of Z's action.

4 Bargmann's Superselection Rule

An often mentioned textbook example where a particular implementation of a symmetry group allegedly clashes with the superposition principle, such that a superselection rule results, is Galilei invariant quantum mechanics (e.g. [9]; see also Wightman's review [39]). We will discuss this example in detail for the general multi-particle case. (Textbook discussions usually restrict to one particle, which, due to Galilei invariance, must necessarily be free.) It will serve as a test case to illustrate the argument of the previous chapter and also to formulate our critique. Its physical significance is limited by the fact that the particular feature of the Galilei group that is responsible for the existence of the mass superselection rule ceases to exist if we replace the Galilei group by the Poincaré group (i.e. it is unstable under 'deformations'). But this is not important for our argument.[12] Let now G be the Galilei group, an element of which is parameterized by $(R, \boldsymbol{v}, \boldsymbol{a}, b)$, with R a rotation matrix in $SO(3)$, $\boldsymbol{v}$ the boost velocity, $\boldsymbol{a}$ the spatial translation, and b the time translation. Its laws of multiplication and inversion are respectively given by

$$\begin{aligned} g_1 g_2 &= (R_1, \boldsymbol{v}_1, \boldsymbol{a}_1, b_1)(R_2, \boldsymbol{v}_2, \boldsymbol{a}_2, b_2) \\ &= (R_1 R_2\,,\ \boldsymbol{v}_1 + R_1 \cdot \boldsymbol{v}_2\,,\ a_1 + R_1 \cdot \boldsymbol{a}_2 + \boldsymbol{v}_1 b_2\,,\ b_1 + b_2), \qquad (26) \\ g^{-1} &= (R, \boldsymbol{v}, \boldsymbol{a}, b)^{-1} = (R^{-1},\ -R^{-1} \cdot \boldsymbol{v}\,,\ -R^{-1} \cdot (\boldsymbol{a} - \boldsymbol{v}b)\,,\ -b). \qquad (27) \end{aligned}$$

We consider the Schrödinger equation for a system of n particles of positions $\boldsymbol{x}_i$, masses m_i, mutual distances $r_{ij} := \|\boldsymbol{x}_i - \boldsymbol{x}_j\|$ which interact via a Galilei-invariant potential $V(\{r_{ij}\})$, so that the Hamilton operator becomes $\mathsf{H} = -\hbar^2 \sum_i \frac{\Delta_i}{2m_i} + V$. The Hilbert space is $\mathcal{H} = L^2(\mathbb{R}^{3n}, \mathrm{d}^3\boldsymbol{x}_1 \cdots \mathrm{d}^3\boldsymbol{x}_n)$.

G acts on the space $\{\text{configurations}\} \times \{\text{times}\} \cong \mathbb{R}^{3n+1}$ as follows: Let $g = (R, \boldsymbol{v}, \boldsymbol{a}, b)$, then $g(\{\boldsymbol{x}_i\}, t) := (\{R \cdot \boldsymbol{x}_i + \boldsymbol{v}t + \boldsymbol{a}\}\,,\ t + b)$. Hence G has the obvious

[11] Had we defined the multiplier exponents mod 2π (compare footnote 10) then we would have obtained a $U(1)$ extension, which would suffice so far. But in the next section we will definitively need the $\mathbb{R}$ extension as symmetry group of the extended classical model discussed there.

[12] In General Relativity, where the total mass can be expressed as a surface integral at 'infinity', the issue of mass superselection comes up again; see e.g. [15] and [8].

left action on complex-valued functions on $\mathbb{R}^{3n+1}$: $(g,\psi) \to \psi \circ g^{-1}$. However, these transformations do *not* map solutions of the Schrödinger equations into solutions. But, as is well known, this can be achieved by introducing an $\mathbb{R}^{3n+1}$-dependent phase factor (see e.g. [13] for a general derivation). We set $M = \sum_i m_i$ for the total mass and $\boldsymbol{r}_c = \frac{1}{M}\sum_i m_i \boldsymbol{x}_i$ for the center-of-mass. Then the modified transformation, T_g, which maps solutions (i.e. curves in $\mathcal{H}$) to solutions, is given by

$$\mathsf{T}_g\psi(\{\boldsymbol{x}_i\},t) := \exp\left(\tfrac{\mathrm{i}}{\hbar}M[\boldsymbol{v}\cdot(\boldsymbol{r}_c-\boldsymbol{a}) - \tfrac{1}{2}\boldsymbol{v}^2(t-b)]\right)\,\psi(g^{-1}(\{\boldsymbol{x}_i\},t)). \tag{28}$$

However, due to the modification, these transformations have lost the property to define an action of G, that is, we do *not* have $\mathsf{T}_{g_1}\circ\mathsf{T}_{g_2} = \mathsf{T}_{g_1g_2}$. Rather, a straightforward calculation using (26) and (27) leads to

$$\mathsf{T}_{g_1}\circ\mathsf{T}_{g_2} = \exp(\mathrm{i}\xi(g_1,g_2))\,\mathsf{T}_{g_1g_2}, \tag{29}$$

with non-trivial multiplier exponent

$$\xi(g_1,g_2) = \tfrac{M}{\hbar}(\boldsymbol{v}_1\cdot R_1\cdot\boldsymbol{a}_2 + \tfrac{1}{2}\boldsymbol{v}_1^2 b_2). \tag{30}$$

Although each T_g is a mapping of *curves* in $\mathcal{H}$, it also defines a unitary transformation on $\mathcal{H}$ itself. This is so because the equations of motion define a bijection between solution curves and initial conditions at, say, $t=0$, which allows to translate the map T_g into a unitary map on $\mathcal{H}$, which we call U_g. It is given by

$$\mathsf{U}_g\psi(\{\boldsymbol{x}_i\}) = \exp\left(\tfrac{\mathrm{i}}{\hbar}M[\boldsymbol{v}\cdot(\boldsymbol{r}_c-\boldsymbol{a}) + \tfrac{1}{2}\boldsymbol{v}^2 b]\right)\,\exp(\tfrac{\mathrm{i}}{\hbar}\mathsf{H}b)\psi(\{R^{-1}(\boldsymbol{x}_i-\boldsymbol{a}+\boldsymbol{v}b)\}), \tag{31}$$

and furnishes a ray-representation whose multiplier exponents are given by (30). It is easy to see that the multiplier exponents are non-trivial, i.e., not removable by a redefinition (22). The quickest way to see this is as follows: suppose to the contrary that they were trivial and that hence (22) holds with $\xi' \equiv 0$. Trivially, this equation will continue to hold after restriction to any subgroup $G_0 \subset G$. We choose for G_0 the abelian subgroup generated by boosts and space translations, so that the combination $\gamma(g_1)-\gamma(g_1g_2)+\gamma(g_2)$ becomes symmetric in $g_1, g_2 \in G_0$. But the exponent (30) stays obviously asymmetric after restriction to G_0. Hence no cancellation can take place, which contradicts our initial assumption.

The same trick immediately shows that the multiplier exponents are inequivalent for different total masses M. Hence, by the general argument given in the previous chapter, if $\mathcal{H}'$ and $\mathcal{H}''$ correspond to Hilbert spaces of states with different overall masses M' and M'', then the requirement that the Galilei group should act on the set of physical states excludes superpositions of states of different overall mass. This is Bargmann's superselection rule.

I criticize these arguments for the following reason: The dynamical framework that we consider here treats 'mass' as parameter(s) which serves to specify the system. States for different overall masses are states of *different* dynamical systems, to which the superposition principle does not even potentially apply. In order to investigate a possible violation of the superposition principle, we must find a dynamical framework in which states of different overall mass are states of the *same* system; in other words, where mass is a dynamical variable. But if we enlarge our system to one where mass is dynamical, it is not at all obvious that the Galilei group will survive as symmetry group. We will now see that in fact it does not, at least for the simple dynamical extension which we now discuss.

The most simple extension of the classical model is to maintain the Hamiltonian, but now regarded as function on an extended, $6n + 2n$ - dimensional phase space with extra 'momenta' m_i and conjugate generalized 'positions' λ_i. Since the λ_i's do not appear in the Hamiltonian, the m_i's are constants of motion. Hence the equations of motion for the $\boldsymbol{x}_i$'s and their conjugate momenta $\boldsymbol{p}_i$ are unchanged (upon inserting the integration constants m_i) and those of the new positions λ_i are

$$\dot{\lambda}_i(t) = \frac{\partial V}{\partial m_i} - \frac{\boldsymbol{p}_i^2}{2m_i^2}, \tag{32}$$

which, upon inserting the solutions $\{\boldsymbol{x}_i(t), \boldsymbol{p}_i(t)\}$, are solved by quadrature.

Now, the point is that the new Hamiltonian equations of motion do not allow the Galilei group as symmetries anymore. But they do allow the $\mathbb{R}$-extension $\bar{G}$ as symmetries [13]. Its multiplication law is given by (24), with ξ as in (30). The action of $\bar{G}$ on the extended space of {configurations} × {times} is now given by

$$\begin{aligned} \bar{g}(\{\boldsymbol{x}_i\},\{\lambda_i\},t) &= (\theta, R, \boldsymbol{v}, \boldsymbol{a}, b)(\{\boldsymbol{x}_i\},\{\lambda_i\},t) \\ &= (\{R\boldsymbol{x}_i + \boldsymbol{v}t + \boldsymbol{a}\}, \{\lambda_i - (\tfrac{\hbar}{M}\theta + \boldsymbol{v}\cdot R\cdot \boldsymbol{x}_i + \tfrac{1}{2}\boldsymbol{v}^2 t)\}, t + b). \end{aligned} \tag{33}$$

With (24) and (30) it is easy to verify that this defines indeed an action. Hence it also defines an action on curves in the new Hilbert space $\bar{\mathcal{H}} := L^2(R^{4n}, \mathrm{d}^{3n}\boldsymbol{x}\, \mathrm{d}^n\lambda)$, given by

$$\bar{\mathsf{T}}_{\bar{g}}\psi := \psi \circ \bar{g}^{-1}, \tag{34}$$

which already maps solutions of the new Schrödinger equation to solutions, *without* invoking non-trivial phase factors. This is seen as follows: Let

$$\Psi(\{\boldsymbol{x}_i\},\{\lambda_i\},t) \in \bar{\mathcal{H}}$$

and $\Phi(\{\boldsymbol{x}_i\},\{m_i\},t)$ its Fourier transform in the (λ_i, m_i) arguments:

$$\Phi(\{\boldsymbol{x}_i\},\{\lambda_i\},t) = (2\pi\hbar)^{-n/2}\int_{\mathbb{R}^n} \mathrm{d}^n m \, \exp\left[\frac{\mathrm{i}}{\hbar}\sum_{i=1}^{n} m_i\lambda_i\right] \Phi(\{\boldsymbol{x}_i\},\{m_i\},t). \tag{35}$$

For each set of masses $\{m_i\}$ the function $\Phi_{\{m_i\}}(\{\boldsymbol{x}_i\},t) := \Phi(\{\boldsymbol{x}_i\},\{m_i\},t)$ satisfies the original Schrödinger equation. Since (34) does not mix different sets of $\{m_i\}$ it induces a map $\bar{\mathsf{T}}_{\bar{g}}^{\{m_i\}}$ for each such set:

$$\bar{\mathsf{T}}_{\bar{g}}^{\{m_i\}}\Phi_{\{m_i\}}(\{\boldsymbol{x}_i\},t) : = \exp\left[\mathrm{i}\theta + \tfrac{\mathrm{i}}{\hbar}M\left(\boldsymbol{v}\cdot(\boldsymbol{r}_c - \boldsymbol{a}) - \tfrac{1}{2}v^2(t-b)\right)\right] \times \Phi_{\{m_i\}}(g^{-1}(\{\boldsymbol{x}_i\},t)) \qquad (36)$$

Via the Fourier transform (35) we represent $\bar{\mathcal{H}}$ as direct integral of $\mathcal{H}_{\{m_i\}}$'s, each of which isomorphic to our old $\mathcal{H} = L^2(\mathbb{R}^{3n}, \mathrm{d}^3\boldsymbol{x}_1\cdots\mathrm{d}^3\boldsymbol{x}_n)$, and on each of which (36) defines a unitary representation U of $\bar{G}$ the form (25) with U_g the ray-representation (31). This shows how the much simpler transformation law (34) contains the more complicated one (28) upon writing $\bar{\mathcal{H}}$ as a direct integral of vector spaces $\mathcal{H}_{\{m_i\}}$.

In the new framework the overall mass, M, is a dynamical variable, and it would make sense to state a superselection rule with respect to it. But now $\bar{G}$ rather than G is the dynamical symmetry group, which acts by a proper unitary representation on $\bar{\mathcal{H}}$, so that the requirement that the dynamical symmetry group should act on the space of physical states will now not lead to any superselection rule. Rather, the new and more physical interpretation of a possible superselection rule for M would be that we cannot localize the system in the coordinate conjugate to overall mass, which we call Λ, i.e., that only the *relative* new positions $\lambda_i - \lambda_j$ are observable.[13] (This is so because M generates translations of equal amount in all λ_i.) But this would now be a contingent physical property rather than a mathematical necessity. Note also that in our dynamical setup it is inconsistent to just state that M generates gauge symmetries, i.e. that Λ corresponds to a physically non existent degree of freedom. For example, a motion in real time along Λ requires a non-vanishing action (for non-vanishing M), due to the term $\int \mathrm{d}t\, M\dot{\Lambda}$ in the expression for the action.

If decoherence were to explain the (ficticious) mass superselection rule, it would be due to a dynamical instability (as explained in [24]) of those states which are more or less localized in Λ. Mathematically this effect would be modelled by effectively removing the projectors onto Λ-subintervalls from the algebra of observables, thereby putting M (i.e. its projectors) into the center of $\mathcal{O}$. Such a non-trivial center should therefore be thought of as resulting from an approximation-dependent idealisation.

[13] A system $\{(\tilde{\lambda}_i, \tilde{m}_i\})$ of canonical coordinates including $M = \sum_i m_i$ is e.g. $\tilde{\lambda}_1 := \lambda_1$, $\tilde{m}_1 = M$ and $\tilde{\lambda}_i = \lambda_i - \lambda_1$, $\tilde{m}_i = m_i$ for $i = 2...n$. Then $\Lambda = \tilde{\lambda}_1$.

5 Charge Superselection Rule

In the previous case I said that superselection rules should be stated within a dynamical framework including as dynamical degree of freedom the direction generated by the superselected quantity. What is this degree of freedom in the case of a superselected electric charge and how does it naturally appear within the dynamical setup? What is its relation to the Coulomb field whose rôle in charge-decoherence has been suggested in [15]? In the following discussion I wish to investigate into these questions by looking at the Hamiltonian formulation of Maxwell's equation and the associated canonical quantization.

In Minkowski space, with preferred coordinates $\{x^\mu = (t,x,y,z)\}$ (laboratory rest frame), we consider the spatially finite region $Z = \{(t,x,y,z) : x^2+y^2+z^2 \le R^2\}$. Σ denotes the intersection of Z with a slice $t =$ const. and $\partial\Sigma =: S_R$ its boundary (the laboratory walls). Suppose we wish to solve Maxwell's equations within Z, allowing for charged solutions. It is well known that in order for charged configurations to be stationary points of the action, the standard action functional has to be supplemented by certain surface terms (see e.g. [11]) which involve new fields on the boundary, which we call λ and f, and which represent a pair of canonically conjugate variables in the Hamiltonian sense. On the laboratory walls, $\partial\Sigma$, we put the boundary conditions that the normal component of the current and the tangential components of the magnetic field vanish. Then the appropriate boundary term for the action reads

$$\int_Z \mathrm{d}t\, \mathrm{d}\omega(\dot{\lambda}+\phi)f, \tag{37}$$

where ϕ is the scalar potential and $d\omega$ the measure on the spatial boundary 2-sphere rescaled to unit radius. Adding this to the standard action functional and expressing all fields on the spatial boundary by their multipole moments (so that integrals $\int_{\partial\Sigma} \mathrm{d}\omega\, R^2$, $\mathrm{d}\omega =$ measure on unit sphere, become $\sum_{lm}$), one arrives at a Hamiltonian function

$$H = \int_\Sigma \left[\tfrac{1}{2}(\boldsymbol{E}^2 + (\boldsymbol{\nabla}\times\boldsymbol{A})^2) + \phi(\rho - \boldsymbol{\nabla}\cdot\boldsymbol{E}) - \boldsymbol{A}\cdot\boldsymbol{j}\right] + \sum_{lm} \phi_{lm}(E_{lm} - f_{lm}). \tag{38}$$

Here the pairs of canonically conjugate variables are $(\boldsymbol{A}(\boldsymbol{x}), -\boldsymbol{E}(\boldsymbol{x}))$ and (λ_{lm}, f_{lm}), and E_{lm} are the multipole components of $\boldsymbol{n}\cdot\boldsymbol{E}$,

$$E_{lm} := \int_{\partial\Sigma} \mathrm{d}\omega R^2\, Y_{lm}\boldsymbol{n}\cdot\boldsymbol{E}, \tag{39}$$

where $\boldsymbol{n}$ is the normal to $\partial\Sigma$. The scalar potential ϕ has to be considered as Lagrange multiplier. With the given boundary conditions the Hamiltonian is differentiable with respect to all the canonical variables[14] and leads to the

[14] This would not be true without the additional surface term (37). Without it one does not simply obtain the wrong Hamiltonian equations of motions, but

following equations of motion

$$\dot{\boldsymbol{A}} = \frac{\delta H}{\delta(-\boldsymbol{E})} = -\boldsymbol{E} - \boldsymbol{\nabla}\phi\,, \tag{40}$$

$$-\dot{\boldsymbol{E}} = -\frac{\delta H}{\delta \boldsymbol{A}} = \boldsymbol{j} - \boldsymbol{\nabla}\times(\boldsymbol{\nabla}\times\boldsymbol{A})\,, \tag{41}$$

$$\dot{\lambda}_{lm} = \frac{\partial H}{\partial f_{lm}} = -\phi_{lm}\,, \tag{42}$$

$$\dot{f}_{lm} = -\frac{\partial H}{\partial \lambda_{lm}} = 0\,. \tag{43}$$

These are supplemented by the equations which one obtains by varying with respect to the scalar potential ϕ, which, as already said, is considered as Lagrange multiplier. Varying first with respect to $\phi(\boldsymbol{x})$ (i.e. within Σ) and then with respect to ϕ_{lm} (i.e. on the boundary $\partial\Sigma$), one obtains

$$G(\boldsymbol{x}) : = \boldsymbol{\nabla}\cdot\boldsymbol{E}(\boldsymbol{x}) - \rho(\boldsymbol{x}) = 0, \tag{44}$$

$$G_{lm} : = E_{lm} - f_{lm} = 0. \tag{45}$$

These equations are constraints (containing no time derivatives) which, once imposed on initial conditions, continue to hold due to the equations of motion. [15]

This ends our discussion of the classical theory. The point was to show that it leaves no ambiguity as to what its dynamical degrees of freedom are, and that we had to include the variables λ_{lm} along with their conjugate momenta f_{lm} in order to gain consistency with the existence of charged configurations. The physical interpretation of the λ_{lm}'s is not obvious. Equation (42) merely relates their time derivative to the scalar potential's multipole moments on the boundary, which are clearly highly non-local quantities. The interpretation of the f_{lm}'s follow from (45) and the definition of E_{lm}, i.e. they are the multipole moments of the electric flux distribution $\varphi(\boldsymbol{n}) := R^2\boldsymbol{n}\cdot\boldsymbol{E}(R^2\boldsymbol{n})$. In particular, for $l = 0 = m$ we have

$$f_{00} = (4\pi)^{-\frac{1}{2}}\,Q, \tag{46}$$

---

none at all! Concerning the Langrangean formalism one should be aware that the Euler-Lagrange equations may formally admit solutions (e.g. with long-ranged (charged) fields) which are outside the class of functions which one used in the variational principle of the action (e.g. rapid fall-off). Such solutions are not stationary points of the action and their admittance is in conflict with the variational principle unless the expression for the action is modified by appropriate boundary terms.

[15] Equation (41) together with charge conservation, $\dot{\rho} + \boldsymbol{\nabla}\cdot\boldsymbol{j} = 0$, shows that (44) is preserved in time, and (43) together with the boundary condition that $\boldsymbol{n}\cdot\boldsymbol{j}$ and $\boldsymbol{n}\times(\boldsymbol{\nabla}\times\boldsymbol{A})$ vanish on $\partial\Sigma$ show that (45) is preserved in time.

where Q is the total charge of the system. Hence we see that the total charge generates motions in λ_{00}. But this means that the degree of freedom labelled by λ_{00} truly exists (in the sense of the theory). For example, a motion along λ_{00} will cost a non-vanishing amount of action $\propto Q(\lambda_{00}^{\text{final}} - \lambda_{00}^{\text{initial}})$. A declaration that λ_{00} really labels only a gauge degree of freedom is *incompatible* with the inclusion of charged states. Similar considerations apply of course to the other values of l, m. But note that this conclusion is independent of the radius R of the spatial boundary 2-sphere $\partial\Sigma$. In particular, it continues to hold in the limit $R \to \infty$. We will not consistently get rid of physical degrees of freedom that way, even if we agree that realistic physical measurements will only detect field values in bounded regions of space-time. See [12] for more discussion on this point and the distinction between proper symmetries and gauge symmetries.

It should be obvious how these last remarks apply to the statement of a charge superselection rule. Without entering the technical issues (see e.g. [33]), its basic ingredient is Gauss' law (for operator-valued quantities), locality of the electric field and causality. That Q commutes with all (quasi-) local observables then follows simply from writing Q as surface integral of the local flux operator $R^2\boldsymbol{n} \cdot \hat{\boldsymbol{E}}$, and the observation that the surface may be taken to lie in the causal complement of any bounded space-time region. Causality then implies commutativity with any local observable.

In a heuristic Schrödinger picture formulation of QED one represents states Ψ by functions of the configuration variables $\boldsymbol{A}(\boldsymbol{x})$ and λ_{lm}. The momentum operators are obtained as usual:

$$-\boldsymbol{E}(\boldsymbol{x}) \longrightarrow -\mathrm{i}\frac{\delta}{\delta \boldsymbol{A}(\boldsymbol{x})}\,, \tag{47}$$

$$f_{lm} \longrightarrow -\mathrm{i}\frac{\partial}{\partial \lambda_{lm}}\,. \tag{48}$$

In particular, the constraint (45) implies the statement that on physical states Ψ we have[16]

$$\hat{Q}\Psi = -\mathrm{i}\sqrt{4\pi}\frac{\partial}{\partial \lambda_{00}}\Psi\,. \tag{49}$$

This shows that a charge superselection rule is equivalent to the statement that we cannot localize the system in its λ_{00} degree of freedom. Removing *by hand* the multiplication operator λ_{00} (i.e. the projectors onto λ_{00}-intervals) from our observables clearly makes Q a central element in the remaining algebra of observables. But what is the physical justification for this removal? Certainly, it is valid FAPP if one restricts to local observations in space-time. To state that this is a *fundamental* restriction, and not only an approximate

[16] Clearly all sorts of points are simply sketched over here. For example, charge quantization presumably means that λ_{00} should be taken with a compact range, which in turn will modify (48) and (49). But this is irrelevant to the point stressed here.

one, is equivalent to saying that for some fundamental reason we cannot have access to some of the *existing* degrees of freedom, which seems at odds with the dynamical setup. Rather, there should be a *dynamical* reason for why localizations in λ_{00} seem FAPP out of reach. The idea of decoherence would be that localizations in λ_{00} are highly unstable against dynamical decoherence.

We have mainly focussed on the charge superselection operator f_{00}, although the foregoing considerations make it clear that by the same argument any two different asymptotic flux distributions also define different superselection sectors of the theory. Do we expect these additional superselection rules to be physically real? First note that for $l > 0$ the f_{lm} are not directly related to the multipole moments of the charge distributions, as the latter fall-off faster than $\frac{1}{r^2}$ and are hence not detectable on the sphere at infinity. Conversely, the higher multipole moments f_{lm} are not measurable (in terms of electromagnetic fields) within any finite region of space-time, unlike the charge, which is tight to massive particles; any finite sphere enclosing all sources has the same total flux. But the f_{lm} can be related to the kinematical state of a particle through the retarded Coulomb field. In fact, given a particle with constant momentum $\boldsymbol{p}$, charge e and mass m, one obtains for the electric flux distribution at time t on a sphere centered at the instantaneous (i.e. at time t) particle position:[17]

$$\varphi_{\boldsymbol{p}}(\boldsymbol{n}) = \frac{em^2}{4\pi}\frac{[\boldsymbol{p}^2+m^2]^{\frac{1}{2}}}{[(\boldsymbol{p}\cdot\boldsymbol{n})^2+m^2]^{\frac{3}{2}}}. \tag{51}$$

Hence different incoming momenta would induce different flux distributions and therefore lie in different sectors. Given that these sectors exist this means

[17] Formula (51) requires a little more explanation: for a particle with general trajectory $\boldsymbol{z}(t)$ let t' be the retarded time for the space-time point $(\boldsymbol{x}, t)$, i.e., $t' = t - \|\boldsymbol{x} - z(t')\|$ ($c = 1$ in our units). Now we can use the well known formula for the retarded electric field (e.g. (14.14) in [21]) and compute the flux distribution on a sphere which lies in the space of constant time t, where it is centered at the retarded position $\boldsymbol{z}(t')$ of the particle. This flux distribution can be expressed as function of the retarded momentum $\boldsymbol{p}' := \boldsymbol{p}(t')$ and the retarded direction $\boldsymbol{n}' := [\boldsymbol{x} - \boldsymbol{z}(t')]/\|\boldsymbol{x} - \boldsymbol{z}(t')\|$ as follows ($E' := \sqrt{\boldsymbol{p}'^2 + m^2}$):

$$\varphi'_{\boldsymbol{p}'}(\boldsymbol{n}') = \frac{em^2}{4\pi}\frac{1}{[E' - \boldsymbol{p}'\cdot\boldsymbol{n}']^2}. \tag{50}$$

If the particle moves with *constant* velocity $\boldsymbol{v} := \dot{\boldsymbol{z}}$, the expression for the retarded Coulomb field can be rewritten in terms of the instantaneous position $\boldsymbol{z}(t)$ by using $\boldsymbol{z}(t) = \boldsymbol{z}(t') + \boldsymbol{v}\|\boldsymbol{x} - \boldsymbol{z}(t')\|$. With respect to this center it is purely radial. Then one calculates the flux distribution on a sphere which again lies in the space of constant time t, but now centered at $\boldsymbol{z}(t)$ rather than $\boldsymbol{z}(t')$. This function can be expressed in terms of the instantaneous direction $\boldsymbol{n} := [\boldsymbol{x} - \boldsymbol{z}(t)]/\|\boldsymbol{x} - \boldsymbol{z}(t)\|$ and the instantaneous momentum $\boldsymbol{p} := \boldsymbol{p}(t)$. One obtains (51).

that different incoming momenta cannot be coherently superposed and no incoming localized states be formed, unless one also adds the appropriate incoming infrared photons to just cancel the difference of the asymptotic flux distributions. This is achieved by imposing the 'infrared coherence condition of Zwanziger [41][18] the effect of which is to 'dress' the charged particles with infrared photons which just subtract their retarded Coulomb fields at large spatial distances. Hence coherent superpositions of particles with different momenta can only be formed if they are dressed by the right amount of incoming infrared photons.

As a technical aside we remark that this can be done without violating the Gupta-Bleuler transversality condition $k^\mu a_\mu(k)|\mathrm{in}\rangle = 0$ in the zero-frequency limit, precisely because of the surface term (37)[11]. This resolved an old issue concerning the compatibility of the infrared coherence condition on one hand, and the Gupta-Bleuler transversality condition on the other [17,42]. From what we said earlier concerning the consistency of the variational principle in the presence of charged states, such an apparent clash of these two conditions had to be expected: without the surface variables one cannot maintain gauge invariance at spatial infinity (i.e. in the infrared limit) and at the same time include charged states. In the charged sectors the longitudinal infrared photons correspond to real physical degrees of freedom and it will naturally lead to inconsistencies if one tries to eliminate them by imposing the Gupta-Bleuler transversality condition also in the infrared limit. However, a gauge symmetry in the infrared limit can be maintained if one adds the asymptotic degrees of freedom in the form of surface terms.

These remarks illustrate how the rich superselection structure associated with different asymptotic flux distributions f_{lm} renders the problem of characterizing state spaces in QED for charged sectors fairly complicated. This problem has been studied within various formalisms including algebraic QFT [5] and lattice approximations, where the algebra of observables can be explicitly presented [25]. However, all this takes for granted the existence of the superselection rules, whereas we would like to see whether they really arise from some physical impossibility to localize the system in the degrees of freedom labelled by λ_{lm}. What physics should prevent us from forming incoming localized wave packets of charged *undressed* (in the sense above) particles, which would produce coherent superpositions of asymptotic flux distributions from the sectors with $l \geq 1$? This cries out for a decoherence mechanism to provide a satisfying physical explanation. The case of charge superselection is, however, more elusive, since localizations in λ_{00} do not have an obvious physical interpretation. Compare the controversy between [1,29] on one side and [36] on the other.

[18] Basically it says that the incoming scattering states should be eigenstates to the photon annihilation operators $a_\mu^{\mathrm{in}}(k)$ in the zero-frequency limit.

References

1. Aharonov Y., Susskind L. (1967): Charge Superselection Rule. Phys. Rev. **155**, 1428-1431
2. Araki H. (1980): A Remark on the Machida-Namiki Theory of Measurement. Prog. Theo. Phys. **64**, 719-730
3. Bargmann V. (1964): Note on Wigner's Theorem on Symmetry Operations. Jour. Math. Phys. **5**, 862-868
4. Bogolubov N.N., Logunov A.A., Oksak A.I., Todorov I.T. (1990): *General Principles of Quantum Field Theory*, (Kluwer, Dordrecht).
5. Buchholz D. (1982): The Physical State Space of Quantum Electrodynamics. Commun. Math. Phys. **85**, 49-71
6. Dirac P.A.M. (1930): *The Principles of Quantum Mechanics* (Clarendon Press, Oxford)
7. Dixmier J. (1981): *Von Neumann Algebras* (North Holland, Amsterdam)
8. Dominguez A.E., Kozameh C.N., Ludvigsen M. (1997): Superselection Sectors in Asymptotic Quantization of Gravity, gr-qc/9609071
9. Galindo A., Pascual P. (1990): *Quantum Mechanics I* (Springer, Berlin)
10. Galindo A., Morales A., Nuñez-Lagos R. (1962): Superselection Principle and Pure States of n-Identical Particles. Jour. Math. Phys. **3**, 324-328
11. Gervais J.L., Zwanziger D. (1980): Derivation From First Principles of the Infrared Structure of Quantum Electrodynamics. Phys. Lett. B **94**, 389-393
12. Giulini D. (1995): Asymptotic Symmetry Groups of Long-Ranged Gauge Configurations. Mod. Phys. Lett A **10**, 2059–2070
13. Giulini D. (1996): On Galilei Invariance and the Bargmann Superselection Rule. *Ann. Phys. (NY)* **249**, 222-235
14. Giulini D. (1995): Quantum Mechanics on Spaces with Finite Fundamental Group. Helv. Phys. Acta **68**, 438–469
15. Giulini D., Kiefer C., Zeh H.D. (1995): Symmetries, Superselection Rules, and Decoherence. Phys. Lett. A **199**, 291–298
16. Giulini D, Joos E., Kiefer C., Kupsch J., Stamatescu I.-O., Zeh H.-D. (1996): *Decoherence and the Appearance of a Classical World in Quantum Theory*, (Springer, Berlin)
17. Haller K. (1978): Gupta-Bleuler condition and infrared-coherent states. Phys.Rev. D **18**, 3045–3051
18. Hartle J.B., Taylor J.R. (1969): Quantum Mechanics of Paraparticles. Phys. Rev. **178**, 2043-2051
19. Hepp K. (1972): Quantum Theory of Measurement and Macroscopic Observables. Helv. Phys. Acta **45**, 237-248
20. Isham C.J. (1995): *Lectures on Quantum Theory.* (Imperial College Press)
21. Jackson J.D. (1975) *Classical Electrodynamics*, second edition, (John Wiley & Sons, New York)
22. Jauch J.M. (1960): Systems of Observables in Quantum Mechanics. Helv. Phys. Acta **33**, 711–726
23. Jauch J.M., Misra B. (1961): Supersymmetries and Essential Observables. Helv. Phys. Acta **34**, 699–709
24. Joos E. (2000): Elements of Environmental Decoherence. In:*Decoherence: Theoretical, Experimental and Conceptual Problems*, Lecture Notes in Physics Vol. 538 (Springer, Berlin), eds. P. Blanchard et al.

25. Kijowski J., Rudolph G., Thielmann A. (1997): Algebra of Observables and Charge Superselection Sectors for QED on the Lattice. Commun. Math. Phys. **188**, 535-564
26. Kupsch J. (2000): Mathematical Aspects of Decoherence. In: *Decoherence: Theoretical, Experimental and Conceptual Problems*, Lecture Notes in Physics Vol. 538(Springer, Berlin), eds. P. Blanchard et al.
27. Landsman N.P. (1995): Observation and Superselection in Quantum Mechanics. Stud. Hist. Phil. Mod. Phys. **26**, 45-73
28. Messiah A.M., Greenberg O.W. (1964): Symmetrization Postulate and Its Experimental Foundation. Phys. Rev. **236**, B 248-B 267
29. Mirman R. (1969): Coherent Superposition of Charge States. Phys. Rev. **186**, 1380–1383
30. Pfeifer P. (1980): A simple model for irreversible dynamics from unitary time evolution. Helv. Phys. Acta **53**, 410-415
31. Primas H. (2000): Asymptotically disjoint quantum states. In: *Decoherence: Theoretical, Experimental, and Conceptual Problems*, Lecture Notes in Physics Vol. 538 (Springer, Berlin), eds. P. Blanchard et al.
32. Raghunathan M.S. (1994): Universal Central Extensions. Rev. Math. Phys. **6**, 207–225
33. Strocchi F., Wightman A.S. (1974): Proof of the Charge Superselection Rule in Local Relativistic Quantum Field Theory. Jour. Math. Phys. **15**, 2198-2224; Erratum Ibid **17** (1976), 1930-1931
34. von Neumann J. (1931): Über Funktionen von Funktionaloperatoren. Ann. Math. (Princeton) **32** 191-226
35. Wick G.C., Wightman A.S., Wigner E.P. (1952): The Intrinsic Parity of Elementary Particles. Phys. Rev. **88**, 101–105
36. Wick G.C., Wightman A.S., Wigner E.P. (1970): Superselection Rule for Charge. Phys. Rev. D **1** 3267–3269
37. Wightman, A.S. (1959): Relativistic Invariance and Quantum Mechanics (Notes by A. Barut). Nouvo Cimento, Suppl. **14**, 81-94
38. Wightman A.S., Glance N. (1989): Superselection Rules in Molecules. Nucl. Phys. B (Proc. Suppl.) **6**, 202-206
39. Wightman A.S. (1995): Superselection Rules; Old and New. Il Nuovo Cimento **110 B**, 751–769
40. Zuerk W. (1982): Environment-Induced Superselection Rules. Phys. Rev. D **26**, 1862-1880
41. Zwanziger D. (1976): Physical States in Quantum Electrodynamics. Phys. Rev. D **14**, 2570-2589
42. Zwanziger D. (1978): Gupta-Bleuler and Infrared-Coherence Subsidiary Conditions. Phys. Rev. D **18**, 3051-3057

Quantum Histories and Their Implications

Adrian Kent

Department of Applied Mathematics and Theoretical Physics, University of Cambridge, Silver Street, Cambridge CB3 9EW, U.K.

Abstract. Classical mechanics and standard Copenhagen quantum mechanics respect subspace implications. For example, if a particle is confined in a particular region R of space, then in these theories we can deduce that it is confined in regions containing R. However, subspace implications are generally violated by versions of quantum theory that assign probabilities to histories, such as the consistent histories approach. I define here a new criterion, ordered consistency, which refines the criterion of consistency and has the property that inferences made by ordered consistent sets do not violate subspace relations. This raises the question: do the operators defining our observations form an ordered consistent history? If so, ordered consistency defines a version of quantum theory with greater predictive power than the consistent histories formalism. If not, and our observations are defined by a non-ordered consistent quantum history, then subspace implications are not generally valid.

1 Introduction

We take it for granted that we can infer quantitatively less precise statements from our observations. For example, if we know that an atom is confined in some region R of space, we believe we are free to assume for calculational purposes only that it lies in some larger region containing R. Our understanding of the world and our interpretation of everyday experience tacitly rely on subspace implications of this general type: if a physical quantity is known to lie within a range R_1, then it lies in all ranges $R_2 \supset R_1$.

In classical mechanics, subspace inferences follow from the correspondence between physical states and points in phase space: if the state of a system lies in a subset S_1 of phase space, it lies in all subsets $S_2 \supset S_1$. Similarly, in Copenhagen quantum theory, they hold since if the state of a quantum system lies in a subspace H_1 of Hilbert space, it lies in all subspaces $H_2 \supset H_1$.

However, neither classical mechanics nor (presumably) Copenhagen quantum theory is fundamentally correct. If the basic principles of quantum theory apply to the universe as a whole, then a post-Copenhagen interpretation of quantum theory seems to be needed, and any justification of the subspace implications must ultimately be given in terms of that interpretation. Though it may seem hard to imagine how to make sense of nature without allowing subspace inferences, there are versions of quantum theory in which they do not hold. In particular, this is true of recent attempts to develop history-based formulations of quantum theory [1–5], which rely on the notion of consistent or decoherent sets of histories.

This paper suggests a way in which quantum theory can plausibly be interpreted via statements about histories, without violating subspace implications — the motivation being that both history-based interpretations and subspace implications seem natural. The interpretation which results, a refinement of the consistent histories approach based on ordered consistent sets of histories, is certainly not the ultimate answer to the problems of quantum theory. It does not solve the measurement problem, for example. But perhaps it is a step in the right direction, or at least in an interesting direction. It also helps to make precise the question as to what it would mean if subspace implications actually *were* violated in nature, which the last part of the paper examines.

The language of quantum histories may not necessarily be the right way to interpret the quantum theory of closed systems. Bohm theory [6] and dynamical collapse theories [7] are particularly interesting alternatives, for example. Quantum histories still play a rôle in these theories as usually interpreted, but it is certainly possible to interpret them so as to ascribe no ontological status to the evolving quantum state, in which case quantum histories have no ontological status either, and the orderings amongst them become physically irrelevant.

In this paper, which focusses on the relation between quantum histories and orderings, I take it for granted — as an interesting hypothesis, not a dogma — that quantum histories have some ontological status. I will focus on the consistent histories approach. This is not to suggest that there are no other interesting quantum histories approaches. Other ideas are outlined in Ref. [8], Section VII of Ref. [9], and Ref. [10], for example. But the discussion of this paper is best carried out within some specific formulation of quantum theory, and consistent histories is the most widely studied example. As an approach to closed system quantum histories, it has many interesting features. Its main drawback is that it does not solve the measurement problem, which in the language of consistent histories takes the form of a set selection problem. The search for ways to refine the definition of consistency in order to solve, or at least reduce, the set selection problem is an independent motivation for the definition of ordered consistency proposed below.

2 Partial Ordering of Quantum Histories

We consider versions of quantum theory that assign probabilities to *histories* — i.e., to collections of events. There are several reasonably standard ways of representing an event in quantum theory. The simplest is as a projection operator, labelled by a particular time, corresponding to a statement about an observable in non-relativistic quantum mechanics. Thus, the statement that a particle was in the interval I at time t is represented by the projection

$$P_I = \int_{x \in I} |x\rangle\langle x| \mathrm{d}^3 x \,. \tag{1}$$

The representation of events as projections can be generalised to *quantum effects* [10–12], defined by operators A such that A and $I - A$ are both positive.[1] Events can also be defined, at least formally, in the path integral version of quantum theory by dividing the set of paths into exclusive subsets. For example, by considering the appropriate integrals one can attach probabilities to the events that a particle did, or did not, enter a particular region of space-time [13]. Further generalisations have been discussed by Isham, Linden and collaborators [5,14,15].

Each of these representations has a natural partial ordering. For projections, we take $A \geq B$ if the range of B is a subspace of that of A. This corresponds to the logical implication of Copenhagen quantum theory already mentioned: if the state of a quantum system lies in the range of B, then it necessarily lies in the range of A. For quantum effects, we take $A \geq B$ if and only if $(A - B)$ is positive. For events defined by the position space path integral, we take the statement that a particle entered R_1 to imply the statement that it entered R_2 if $R_1 \subseteq R_2$.

We define a quantum history as a collection of quantum events and extend the partial ordering to histories in the natural way. For example, if we take quantum histories in non-relativistic quantum mechanics to be defined by sequences of projections at different fixed times, we compare two quantum histories $H = \{A_1, t_1; \ldots ; A_n; t_n\}$ and $H' = \{A'_1, t'_1; \ldots ; A'_{n'}, t'_{n'}\}$ as follows. First add to each history the identity operator at every time at which it contains no proposition and the other does, so obtaining relabelled representations of the histories as $H = \{A_1, T_1; \ldots ; A_N; T_N\}$ and $H' = \{A'_1, T_1; \ldots ; A'_N, T_N\}$. Then define the partial ordering by taking $H \geq H'$ if and only if $A_i \geq A'_i$ for all i and $H > H'$ if $H \geq H'$ and $H \neq H'$. This ordering was first considered in the consistent histories literature by Isham and Linden [14], whose work, and its relation to the ideas of this paper, are discussed in the Appendix.

3 Consistent Histories

The consistent sets of histories for a closed quantum system are defined in terms of the space of states $\mathcal{H}$, the initial density matrix ρ, and the hamiltonian H. In the simplest version of the consistent histories formulation of non-relativistic quantum mechanics, sets of histories correspond to sets of projective decompositions. In order to be able to give a physical interpretation of any of the consistent sets, we need also to assume that standard observables, such as position, momentum and spin, are given. Individual quantum

[1] Different definitions can be found: this, the simplest, is adequate for our purposes.

events are defined by members of projective decompositions of the identity into orthogonal hermitian projections $\sigma = \{P^i\}$, with

$$\sum_i P^i = 1 \quad \text{and} \quad P^i P^j = \delta_{ij} P^i \,. \tag{2}$$

Each such decomposition defines a complete and exclusive list of events at some fixed time, and a time label is thus generally attached to the decompositions and the projections: the labels are omitted here, since the properties of interest here depend only on the time ordering of events.

Suppose now we have a set of decompositions $\mathcal{S} = \{\sigma_1, \ldots, \sigma_n\}$. Then the histories given by choosing one projection from each of the decompositions σ_j in all possible ways define an exhaustive and exclusive set of alternatives. We follow Gell-Mann and Hartle's definitions, and say that $\mathcal{S}$ is a *consistent* (or *medium decoherent*) set of histories if

$$\mathrm{Tr}(P_n^{i_n} \ldots P_1^{i_1} \rho P_1^{j_1} \ldots P_n^{j_n}) = \delta_{i_1 j_1} \ldots \delta_{i_n j_n} p(i_1 \ldots i_n) \,. \tag{3}$$

When $\mathcal{S}$ is consistent, $p(i_1 \ldots i_n)$ is the probability of the history $H = \{P_1^{i_1}, \ldots, P_n^{i_n}\}$. We will want later to discuss the properties of individual histories without reference to any fixed consistent set, and we define a *consistent history* to be a history which belongs to some consistent set $\mathcal{S}$. Finally, we say the set

$$\mathcal{S}' = \{\sigma_1, \ldots, \sigma_k, \tau, \sigma_{k+1}, \ldots, \sigma_n\} \tag{4}$$

is a *consistent extension* of a consistent set of histories $\mathcal{S} = \{\sigma_1, \ldots, \sigma_n\}$ by the set of projections $\tau = \{Q^i : i = 1, \ldots, m\}$ if τ is a projective decomposition and $\mathcal{S}'$ is consistent.

Suppose now that we have a collection of data defined by the history

$$H = \{P_1^{i_1}, \ldots, P_n^{i_n}\} \tag{5}$$

which has non-zero probability and belongs to the consistent set $\mathcal{S}$. This history might, for example, describe the results of a series of experiments or the observations made by an observer. Given a choice of consistent extension $\mathcal{S}'$ of $\mathcal{S}$, we can make probabilistic inferences conditioned on the history H. For example, if $\mathcal{S}'$ has the above form, the histories extending H in $\mathcal{S}'$ are $H^i = \{P_1^{i_1}, \ldots, P_k^{i_k}, Q^i, P_{k+1}^{i_{k+1}}, \ldots, P_n^{i_n}\}$ and the proposition Q^i has conditional probability

$$p(Q^i | H) = p(H^i)/p(H) \,. \tag{6}$$

These probabilities and conditional probabilities are the same in every consistent set which includes H^i and (hence) H. However, when we want to emphasize that the calculation can be carried out in some particular set $\mathcal{S}$, we will attach a suffix. For example, we might write $p_{\mathcal{S}'}(Q^i|H) = p_{\mathcal{S}'}(H^i)/p_{\mathcal{S}}(H)$ for the above equation.

The formalism itself gives no way of choosing any particular consistent extension. In the view of the original developers of the consistent histories approach, the different S' are to be thought of as ways of producing possible pictures of the past and future physics of the system which, though generally incompatible, are all equally valid. More formally, they can be seen as incompatible logical structures which allow different classes of inferences from the data [16,17].

It is this freedom in the choice of consistent extension which, it has been argued elsewhere [9,18–23,10] gives rise to the most serious problems in the consistent histories approach. Standard probabilistic predictions and deterministic retrodictions can be reproduced in the consistent histories formalism by making an ad hoc choice of consistent set, but cannot be derived from the formalism itself. In fact, it is almost never possible to make any unambiguous predictions or retrodictions: there are almost always an infinite number of incompatible consistent extensions of the set containing a given history dataset [9,20].

The problem is not simply that the formalism supplies descriptions of physics which are complementary in the standard sense, although that in itself is sufficient to ensure that the formalism is only very weakly predictive. Even on the assumption that we will continue to observe quasiclassical physics, no known interpretation of the formalism allows us to derive the predictions of classical mechanics and Copenhagen quantum theory [18]. Hence the set selection problem: probabilistic predictions can only be made conditional on a choice of a consistent set, yet the consistent histories formalism gives no way of singling out any particular set or sets as physically interesting.

One possible solution to the set selection problem would be an axiom which identifies a unique physically interesting set, or perhaps a class of such sets, from the initial state and the dynamics. Another would be the identification of a physically natural measure on the space of consistent sets, according to which the physically relevant consistent set is randomly chosen. Possible set selection criteria have been investigated [24,25], but no generally workable criterion has emerged. [2]

4 Consistent Sets and Contrary Inferences: A Brief Review

A further reason for believing that the consistent histories formalism is at best incomplete comes from considering the logical relations among events in different consistent sets. We say that two projection operators P and Q are *complementary* if they do not commute: $PQ \neq QP$. We say that they are *contradictory* if they sum to the identity, so that $P = 1 - Q$ and $PQ = QP =$

[2] With different motivations, Gell-Mann and Hartle have explored a "strong decoherence" criterion which is intended to reduce the number of consistent sets [26]. However, under their present definition, every consistent set is strongly decoherent [10].

0, and that they are *contrary* if they are orthogonal and not contradictory, so that $PQ = QP = 0$ and $P < 1 - Q$.

Contradictory inferences are never possible in the consistent histories formalism, but it is easy to find examples of contrary inferences from the same data [19]. For instance, consider a quantum system whose hamiltonian is zero and whose Hilbert space $\mathcal{H}$ has dimension greater than two, prepared in the state $|a\rangle$. Suppose that the system is left undisturbed from time 0 until time t, when it is observed in the state $|c\rangle$, where $0 < |\langle a | c \rangle| \leq 1/3$, i.e. a single quantum measurement is made, and the outcome probability is less than 1/9. In consistent histories language, we have the initial density matrix $\rho = |a\rangle\langle a|$ and the single datum corresponding to the history $H = \{P_c\}$ from the consistent set $\mathcal{S} = \{\{P_c, 1 - P_c\}\}$, where the projection $P_c = |c\rangle\langle c|$ is taken at time t.

Now consider consistent extensions of $\mathcal{S}$ of the form $\mathcal{S}_b = \{\{P_b, 1 - P_b\}, \{P_c, 1 - P_c\}\}$, where $P_b = |b\rangle\langle b|$ for some normalised vector $|b\rangle$ with the property that

$$\langle c | b \rangle \langle b | a \rangle = \langle c | a \rangle . \tag{7}$$

It is easy to verify that $\mathcal{S}_b$ is consistent and that the conditional probability of P_b given H is 1. It is also easy to see that there are at least two mutually orthogonal vectors $|b\rangle$ satisfying (7) . For example, let $|v_1\rangle, |v_2\rangle, |v_3\rangle$ be orthonormal vectors and take $|a\rangle = |v_1\rangle$ and $|c\rangle = \lambda|v_1\rangle + \mu|v_2\rangle$, where $|\lambda|^2 + |\mu|^2 = 1$. Then the vectors

$$|b_{\pm}\rangle = (|\lambda|^2 + \frac{|\mu|^2}{x})^{-1/2}(\lambda|v_1\rangle + \frac{\mu}{x}|v_2\rangle \pm \frac{(x-1)^{1/2}\mu}{x}|v_3\rangle) \tag{8}$$

both satisfy (7) and are orthogonal if x is real and $x^2|\lambda|^2 = (x-2)(1-|\lambda|^2)$, which has solutions for $|\lambda| \leq 1/3$. Thus the consistent sets $\mathcal{S}_{b_{\pm}}$ give contrary probability on retrodictions.

Some brief historical remarks are in order. The existence of contrary inferences in the consistent histories formalism, though easy to show, was noticed only quite recently. In particular, it was not known to the formalism's original developers.[3] It was first explicitly pointed out, and its implications for the consistent histories formalism were first examined, in Ref. [19]. Further discussion can be found in Refs. [27,28]. However, a noteworthy earlier consistent histories analysis of an example in which contrary inferences arise can be found in a critique by Cohen [29] of Aharonov and Vaidman's interpretation [30] of one of their intriguing examples of pre- and post-selection.[4] As noted in Ref. [19], Cohen's analysis miscontrues the consistency criterion: however, this error does not affect its derivation of contrary inferences.

[3] I am grateful to Bob Griffiths, Jim Hartle, and Roland Omnès for helpful correspondence on this point.

[4] I am grateful to Oliver Cohen and Lucien Hardy for drawing this reference to my attention.

Now, the existence of contrary inferences in the consistent histories formalism needs to be interpreted with care. It is *not* true that, in any given consistent set, two different contrary propositions can be inferred with probability one. The inferences made within any given consistent set lead to no contradiction. The picture of physics given by any given consistent set may or may not be considered natural or plausible — depending on one's intuition and the criteria one uses for naturality — but it is not logically self-contradictory. It *is* however true, as a mathematical statement about the properties of the consistent histories formalism, that the propositions inferred in the two different sets correspond to contrary projections. The formalism makes no physical distinction among different consistent sets, and so requires us to conclude that two equally valid pictures of physics can be given, in which contrary events take place.

To put it more formally, the consistent histories approach can be interpreted as setting out rules of reasoning according to which, although physics can be described by any of infinitely many equally valid pictures, only one of those pictures may be considered in any given argument. Such an interpretation ensures — tautologically — that no logical contradiction arises, even when the pictures contain contrary inferences.

The consistent histories formalism, in other words, gives a set of rules for producing possible pictures of physics within quantum theory, and these rules themselves lead to no logical inconsistency. However, consistent historians claim much more, arguing that the formalism defines a natural and scientifically unproblematic interpretation of quantum theory. Indeed, the consistent histories literature tends to suggest that the descriptions of physics given by consistent historians are simply and evidently the correct descriptions which emerge from quantum theory, so that, in querying them, one necessarily queries quantum theory itself.[5]

This seems patently false. The most basic premise of consistent historians — that quantum theory is correctly interpreted by some sort of many-picture scheme — leads to such trouble in explaining which particular picture we see, and why, that cautious scepticism seems only appropriate. Even if the premise were accepted, it would be essential to ask, of any particular many-picture scheme, whether its assumptions are natural and whether the descriptions of nature it produces are physically plausible or scientifically useful. The particular equations used to define consistent sets are, after all, simply interesting guesses: there is no compelling theoretical justification for them, and indeed, several different definitions of consistency have been proposed [1,4,32].

My own view is that there are a number of compelling reasons for regarding the consistent histories interpretation, as it is presently understood, as scientifically unsatisfactory. However, as these questions have been explored

[5] For example, the consistent histories interpretation of quantum mechanics has been referred to as "the interpretation of quantum mechanics"[16] and even as simply "quantum mechanics"[31].

in some detail elsewhere [9,10,18–23,27,28], I here comment only on two specific problems raised by contrary inferences.

First, the fact that we are to take as equally valid and correct pictures of physics which include contrary inferences goes against many well developed intuitions. No argument based on intuition alone can be conclusive, but I think it must be granted that this one has some force. What use, it may reasonably be asked, is there in saying that in one picture of reality a particle genuinely went, with probability one, through slit A, and that in another picture the particle went, also with probability one, through the disjoint slit B? Why should we take either picture seriously, given the other?

On this point, it is worth noting that one of the advertised merits of the consistent histories formalism [1,16,17] is that it, unlike the Copenhagen interpretation, accommodates some (arguably) plausible intuitions about the behaviour of microsystems in between observations. For example, the formalism allows us to say — albeit only as one of an infinite number of incompatible descriptions — that a particle observed at a particular detector was travelling towards that detector before the observation, and that a particle measured to have spin component $\sigma_x = s_x$ had that spin before the measurement took place. As Griffiths and Omnès note [1,16], informal discussions of experiments are often framed in terms which, if taken literally, suggest that we can make this sort of statement about microsystems before a measurement is carried out. ("Was the beam correctly aligned going into the second interferometer?", or "Do you think something crazy in the electronics might have triggered [detector] number 3 just before the particle got there?" [1], for example.) Their intuition is that a good interpretation of quantum theory ought to give a way of allowing us to take such statements literally — a criterion which, they suggest, the consistent histories approach satisfies.

The intuition is, of course, controversial. A counter-intuition, which most interpretations of quantum theory support, is that any description of a microsystem before a measurement is carried out should be independent of the result of that measurement.

In any case, to the extent that any intuition is offered as a justification of the formalism, it seems reasonable to consider the fact that the formalism violates other strongly held intuitions. Few experimenters, after all, can ever have intuitively concluded that the *entire* flux of their beam can sensibly be thought of as having followed any of several macroscopically distinct paths through the apparatus. Yet this is what the above example, translated into an interference experiment, implies.

The second, and probably deeper, problem is that it seems very hard to justify the distinction, which consistent historians are forced to draw, between contradictory inferences, which are regarded as *a priori* unacceptable, and contrary inferences, which are regarded as unproblematic. Some justification seems called for, since the distinction is not an accidental feature: it is not that the formalism, for unrelated reasons, simply happens to exclude one type of inference and include the other. The definition of consistency is motivated precisely by the notion that, when two different sets allow a calculation of the probability of the same event (belonging, in the simplest case, to a single history in one set and a combination of two histories in the other), the calculations should agree. This requires in particular that contradictory propositions P and $(1-P)$ can never be inferred, since if the probability of P is one in any set, it must be one in all sets, and so the probability of $(1-P)$ must be zero in all sets.

Now this last requirement is not absolutely essential, sensible though it may seem. No logical contradiction arises in an interpretation of quantum theory which follows the basic interpretational ideas of the consistent histories formalism but which accepts all complete sets of disjoint quantum histories, whether consistent or not, as defining valid pictures of physics [9,10]. In this "inconsistent histories" interpretation, contradictory inferences can generally be made by using different pictures. This possibility is excluded by a deliberate theoretical choice.

It seems natural, then, having made this choice, to look for ways in which contrary inferences can similarly be excluded. This is the line of thought pursued below. Note, however, that the problem of contrary inferences is not the only motivation for the ideas introduced below. Whatever one's view of the consistent histories formalism, it is interesting that an alternative formalism can be defined relatively simply. It seems fruitful to ask which, if either, is to be preferred, and why. And, as we will see, ordered consistent sets raise independently interesting questions about the quasiclassical world we actually observe.

5 Relation of Contrary Inferences and Subspace Implications

A contrary inference arises when there exist two consistent sets, $\mathcal{S}_1$ and $\mathcal{S}_2$, both containing a history H, with the property that there are orthogonal propositions P_1 and P_2 which are implied by H in the respective sets, so that — temporarily adding set suffices for clarity — we have

$$p_{\mathcal{S}_1}(P_1|H) = p_{\mathcal{S}_2}(P_2|H) = 1\,. \tag{9}$$

Now $p_{\mathcal{S}_2}((1-P_2)|H) = 0$, and since the probabilities are set-independent and $p_{\mathcal{S}_1}(P_1|H)$ is nonzero, we cannot have $P_1 = 1 - P_2$. Hence, since P_1 and P_2 are orthogonal, we have that $P_1 < 1 - P_2$. Since $p(P_1|H) = 1$ and $p((1-P_2)|H) = 0$, this pair of projections violates the subspace implication $P_1 \Rightarrow 1 - P_2$. That is, a contrary inference implies the existence of consistent histories H and H', belonging to different consistent sets and agreeing on all but one projector, such that H has non-zero probability, H' has zero probability, and $H < H'$: in the example of the last section, for instance, we have $H = \{P_{b_+}, P_c\}$ and $H' = \{(1 - P_{b_-}), P_c\}$.

Clearly, according to the consistent histories formalism, an observation of the datum P_{b_+} cannot be taken to imply an observation of the strictly larger projector $(1 - P_{b_-})$. To make that inference would lead directly to a contradiction, in the form of the realisation of a probability zero history, if P_c were subsequently observed.

Now it is easy to produce real world examples of contrary inferences, so long as those inferences are of unobserved quantities. As we have seen, it requires only a three-dimensional quantum system, prepared in one state, isolated, and then observed in another state — hardly a taxing experiment.

It is not obvious, though, that we can produce examples where subspace implications fail in a realistic consistent histories description of *observations* of laboratory experiments, or more generally of macroscopic quasiclassical physics. That general consistent histories violate subspace implications need not necessarily imply that the particular consistent histories used to recover standard descriptions of real world physics do so. Both in order to address this question, and for its own sake, it is interesting to ask whether there might be any alternative treatment of quantum theory within the consistent histories framework which respects subspace implications, at least when they relate two consistent histories. The next section suggests such a treatment.

6 Ordered Consistent Sets of Histories

We have already seen that there is a natural partial ordering for each of the standard representations of quantum histories in the consistent histories approach. The probability weight defines a second partial ordering: $H \prec H'$ if $p(H) < p(H')$. The violation of subspace implications reflects the disagreement between these two partial orderings in the consistent histories formalism: we can have both $H < H'$ and $H \succ H'$. The aim of this section is to develop a history-based interpretation which restricts attention to collections of quantum histories on which the two orderings do not disagree.

We begin at the level of individual quantum histories, defining an *ordered consistent history*, H, to be a consistent history with the properties that:

(i) for all consistent histories H' with $H' \geq H$ we have that $p(H') \geq p(H)$;
(ii) for all consistent histories H' with $H' \leq H$ we have that $p(H') \leq p(H)$.

Recall that a consistent history is any quantum history which belongs to *some* consistent set of histories. Properties (i) and (ii) hold trivially for histories H and H' which belong to the same consistent set: it is the comparison across different sets which makes them useful constraints.

We now define an *ordered consistent set of histories* to be a complete set of exclusive alternative histories, each of which is ordered consistent. We can then define an ordered consistent histories approach to quantum theory in precise analogy to the consistent histories approach, using the same definition of probability weight and the same interpretation, simply declaring by fiat that only ordered consistent sets of histories are to be considered.[6]

The following lemmas show that, within the projection operator formulation, ordered consistent sets of histories do indeed exist.[7]

Lemma 1: Any consistent history $H = \{P_1, \ldots, P_n\}$ defined by a series of projections which include a minimal projection P_j, so that $P_i \geq P_j$ for all i, is an ordered consistent history.

Proof: Suppose $H' = \{P'_1, \ldots, P'_n\}$ is a larger history than H, and write $P'_i = P_j + Q_i$. We have that

$$\begin{aligned} p(H') &= \mathrm{Tr}(P'_n \ldots P'_1 \rho P'_1 \ldots P'_n) \\ &= \mathrm{Tr}((P_j + Q_n) \ldots (P_j + Q_1)\rho(P_j + Q_1) \ldots (P_j + Q_n)) \\ &= \mathrm{Tr}(P_j \rho) + \mathrm{Tr}(Q_n \ldots Q_1 \rho Q_1 \ldots Q_n) \\ &\geq \mathrm{Tr}(P_j \rho) \\ &= p(H) \,. \end{aligned} \tag{10}$$

Now suppose that H' is smaller than H. Then in particular $P'_j \leq P_j$ and we have that

$$\begin{aligned} p(H') &\leq \mathrm{Tr}(P'_j \rho) \\ &\leq \mathrm{Tr}(P_j \rho) \\ &= p(H) \,. \end{aligned} \tag{11}$$

Lemma 2: Let $\mathcal{S} = \{\sigma_1, \ldots, \sigma_n\}$ be a consistent set of histories defined by a series of projective decompositions, of which one, σ_j, has the property that for each projection P in σ_j, and for every i, we have that there is precisely one projection Q in σ_i with the property that $Q \geq P$ (so that all of the other projections in σ_i are contrary to P). Then $\mathcal{S}$ is an ordered consistent set of histories.

Proof: Each of the histories of non-zero probability in $\mathcal{S}$ satisfies the conditions of Lemma 1 and so is ordered consistent. Each of the histories of zero probability in $\mathcal{S}$ is of the form $H = \{\ldots, P, \ldots, Q, \ldots\}$, where P and Q are contrary projections. Now any consistent history smaller than H therefore also contains a pair of contrary projections $P' \leq P$ and $Q' \leq Q$. By the

[6] Note that there are other collections of quantum histories on which the orderings do not disagree. For example, if all the consistent histories H that violate (i) are eliminated, the remainder form a collection on which the orderings do not disagree and which is not obviously identical to the ordered consistent histories, and similarly for (ii). It might be interesting to explore such alternatives, but we restrict attention to the ordered consistent histories here.

[7] I am grateful to Bob Griffiths for suggesting a slight extension of Lemma 1.

consistency axioms, its probability is less than or equal to $\mathrm{Tr}(Q'P'\rho P'Q') = 0$, and thus must also be zero. Hence H is ordered consistent, since any consistent history larger than H has probability greater than or equal to zero.

7 Ordered Consistent Sets and Quasiclassicality

A formalism based on ordered consistent sets of histories obviously defines a more strongly predictive version of quantum theory than that defined by the existing consistent histories framework, since it allows strictly fewer sets as possible descriptions of physics. But can it describe our empirical observations?

The question subdivides. Are quasiclassical domains generally ordered consistent sets of histories? Is our own quasiclassical domain one? If not, are its histories generally ordered when compared to histories belonging to other consistent sets defined by projections onto ranges of the same quasiclassical variables? For example, can we show that consistent histories defined by projections onto ranges of densities for chemical species in small volumes are generally ordered with respect to one another? If so, then the type of subspace implication which is generally used in analyses of observations could still be justified. Finally, if either of the previous two properties fail to hold, it would be useful to quantify the extent to which they fail.

Answering any of these questions definitively may require — and, it might be hoped, help to develop — a deeper understanding of quasiclassicality than is available to us at present. I at any rate do not know the answers, and can only offer the questions as interesting ones whose resolution would have significant implications. At least ordered consistency does not seem to fall at the first hurdle: ordered consistent sets are shown below to be adequate to describe quasiclassicality in simple models.

As a simple example, consider the following model of a series of successive measurements of the spin of a spin-1/2 particle about various axes. We use a vector notation for the particle states, so that if $\mathbf{u}$ is a unit vector in R^3 the eigenstates of $\boldsymbol{\sigma}\cdot\mathbf{u}$ are represented by $|\pm\mathbf{u}\rangle$. With the analogy of a pointer state in mind, we use the basis $\{|\uparrow\rangle_k, |\downarrow\rangle_k\}$ to represent the kth environment particle state, together with the linear combinations $|\pm\rangle_k = (|\uparrow\rangle_k \pm |\downarrow\rangle_k)/\sqrt{2}$. We compactify the notation by writing environment states as single kets, so that for example $|\uparrow\rangle_1 \otimes \cdots \otimes |\uparrow\rangle_n$ is written as $|\uparrow \ldots \uparrow\rangle$, and we take the initial state $|\psi(0)\rangle$ to be $|\mathbf{v}\rangle \otimes |\uparrow \ldots \uparrow\rangle$.

The interaction between the system and the kth environment particle is chosen so that it corresponds to a measurement of the system spin along the $\mathbf{u}_k$ direction, so that the states evolve as follows:

$$\begin{aligned} |\mathbf{u}_k\rangle \otimes |\uparrow\rangle_k &\to |\mathbf{u}_k\rangle \otimes |\uparrow\rangle_k\,, \\ |-\mathbf{u}_k\rangle \otimes |\uparrow\rangle_k &\to |-\mathbf{u}_k\rangle \otimes |\downarrow\rangle_k\,. \end{aligned} \tag{12}$$

A simple unitary operator that generates this evolution is

$$U_k(t) = P(\mathbf{u}_k) \otimes I_k + P(-\mathbf{u}_k) \otimes \exp(-\mathrm{i}\theta_k(t)F_k)\,, \tag{13}$$

where $P(\mathbf{x}) = |\mathbf{x}\rangle\langle\mathbf{x}|$ and $F_k = \mathrm{i}|\downarrow\rangle_k\langle\uparrow|_k - \mathrm{i}|\uparrow\rangle_k\langle\downarrow|_k$. Here $\theta_k(t)$ is a function defined for each particle k, which varies from 0 to $\pi/2$ and represents how far the interaction has progressed. We define $P_k(\pm) = |\pm\rangle_k\langle\pm|_k$, so that $F_k = P_k(+) - P_k(-)$.

The Hamiltonian for this interaction is thus

$$H_k(t) = \mathrm{i}\dot{U}_k(t)U_k^\dagger(t) = \dot{\theta}_k(t)P(-\mathbf{u}_k) \otimes F_k\,, \tag{14}$$

in both the Schrödinger and Heisenberg pictures. We write the extension of U_k to the total Hilbert space as

$$\begin{aligned} V_k = {} & P(\mathbf{u}_k) \otimes I_1 \otimes \cdots \otimes I_n \\ & + P(-\mathbf{u}_k) \otimes I_1 \otimes \cdots \otimes I_{k-1} \otimes \exp(-\mathrm{i}\theta_k(t)F_k) \otimes I_{k+1} \otimes \cdots \otimes I_n\,. \end{aligned} \tag{15}$$

We take the system particle to interact initially with particle 1 and then with consecutively numbered ones, and there is no interaction between environment particles, so that the evolution operator for the complete system is

$$U(t) = V_n(t) \ldots V_1(t)\,, \tag{16}$$

with each factor affecting only the Hilbert spaces of the system and one of the environment spins.

We suppose, finally, that the interactions take place in disjoint time intervals and that the first interaction begins at $t = 0$, so that the total Hamiltonian is simply

$$H(t) = \sum_{k=1}^{n} H_k(t)\,, \tag{17}$$

and we have that $\theta_1(t) > 0$ for $t > 0$ and that, if $0 < \theta_k(t) < \pi/2$, then $\theta_i(t) = \pi/2$ for all $i < k$ and $\theta_i(t) = 0$ for all $i > k$.

This model has been used elsewhere [25,33] in order to explore algorithms which might select a single physically natural consistent set when the physics is determined by the simplest type of system-environment interaction. It is particularly well suited to such an analysis, since the dynamics are chosen so as to allow a simple and quite elegant classification [33] of all the consistent sets built from projections onto subspaces defined by the Schmidt decomposition. Apart from this, though, the model is unexceptional — one of the simpler variants among the many models used in the literature to investigate the decoherence of system states by measurement-type interactions with an environment.

To give a physical interpretation of the model, we take it that the environment "pointer" variables assume definite values after their respective interactions with the system. That is, after the $k^{\rm th}$ interaction, the $k^{\rm th}$ environment particle is in one of the states $|\uparrow\rangle_k$ and $|\downarrow\rangle_k$: the probabilities of each of these outcomes depend on the outcome of the previous measurement (or, in the case of the first measurement, on the initial state) via the standard quantum mechanical expressions.

This description can be recovered from the consistent histories formalism by choosing the consistent set $\mathcal{S}_1$, defined by the decompositions

$$\begin{aligned}
&\{\, I \otimes |\epsilon_1\rangle_1 \langle \epsilon_1|_1 \otimes I \otimes \cdots \otimes I : \epsilon_1 = \uparrow \text{ or } \downarrow\} \text{ at time } t_1 \,, \qquad (18)\\
&\{\, I \otimes |\epsilon_1\rangle_1 \langle \epsilon_1|_1 \otimes |\epsilon_2\rangle_2 \langle \epsilon_2|_2 \otimes \cdots \otimes I : \epsilon_1, \epsilon_2 = \uparrow \text{ or } \downarrow\} \text{ at time } t_2 \,,\\
&\ldots\\
&\{\, I \otimes |\epsilon_1\rangle_1 \langle \epsilon_1|_1 \otimes |\epsilon_2\rangle_2 \langle \epsilon_2|_2 \otimes \cdots \otimes |\epsilon_n\rangle_n \langle \epsilon_n|_n : \epsilon_1, \epsilon_2, \ldots, \epsilon_n = \uparrow \text{ or } \downarrow\}\\
&\qquad\qquad\qquad\qquad\qquad\qquad\qquad\qquad \text{at time } t_n \,.
\end{aligned}$$

Clearly, the histories of non-zero probability in $\mathcal{S}_1$ take the form

$$\begin{aligned}
H_{\epsilon_1,\ldots,\epsilon_n} = \{ & I \otimes I \otimes I \otimes \cdots \otimes I,\\
& I \otimes |\epsilon_1\rangle_1 \langle \epsilon_1|_1 \otimes I \otimes \cdots \otimes I,\\
& I \otimes |\epsilon_1\rangle_1 \langle \epsilon_1|_1 \otimes |\epsilon_2\rangle_2 \langle \epsilon_2|_2 \otimes I \otimes \cdots \otimes I,\\
& \ldots,\\
& I \otimes |\epsilon_1\rangle_1 \langle \epsilon_1|_1 \otimes |\epsilon_2\rangle_2 \langle \epsilon_2|_2 \otimes \cdots \otimes |\epsilon_n\rangle_n \langle \epsilon_n|_n \} \,, \qquad (19)
\end{aligned}$$

for sequences $\{\epsilon_1, \ldots, \epsilon_n\}$, each element of which takes the value $\uparrow$ or $\downarrow$. Their probabilities, defined by the decoherence functional, are precisely those which would be obtained from standard quantum theory by treating each interaction as a measurement:

$$p(H_{\epsilon_1,\ldots,\epsilon_n}) = (\frac{1 + a_1 \mathbf{v}.\mathbf{u_1}}{2})(\frac{1 + a_2 \mathbf{u_1}.\mathbf{u_2}}{2}) \ldots (\frac{1 + a_n \mathbf{u_{n-1}}.\mathbf{u_n}}{2}) \,, \qquad (20)$$

where, letting $\epsilon_0 = \uparrow$, we define $a_i = 1$ if ϵ_i and ϵ_{i-1} take the same value, and $a_i = -1$ otherwise.

Now $\mathcal{S}_1$ is defined by a nested sequence of increasingly refined projective decompositions, all of whose projections commute — a relation which is unaltered by moving to the Heisenberg picture. It therefore satisfies the conditions of Lemma 2 above, and so is ordered consistent.

This argument clearly generalizes: in any situation in which Hilbert space factorizes into system and environment degrees of freedom, where the self-interactions of the latter are negligible, any consistent set defined by nested commuting projections onto the environment variables is ordered consistent. The model considered above is a particularly crude example: more sophisticated, and phenomenologically somewhat more plausible, examples of this type are analysed in, for example, Refs. [26,32].

No sweeping conclusion can be drawn from this, since it is generally agreed that familiar quasiclassical physics is *not* well described in general — at least in any obvious way — by models of this type. (Again, a detailed discussion of the limitations of such models can be found in Refs. [26,32].) In other words, while it would be hard to defend the hypothesis that familiar quasiclassical sets are generally ordered consistent if sets of the type $\mathcal{S}_1$ were not, the fact that they are is certainly not sufficient evidence. It would be good to find sharper tests of the hypothesis, perhaps for example by developing further the phenomenological investigations of quasiclassicality pursued in Ref. [26]. Meanwhile, the questions raised earlier in this section remain unresolved.

On the other hand, it would be difficult to make a watertight case that ordered consistent sets are definitely inadequate to describe real-world physics, for the following reason. First, it seems hard to exclude the possibility that the initial state is pure, so let us temporarily suppose that it is: $\rho = |\psi\rangle\langle\psi|$. As Gell-Mann and Hartle point out [32], we can then associate to every consistent set of histories, $\mathcal{S}$, a nested set of commuting projections defining what they term generalized records. The consistent set defines a resolution of the initial state into history vectors,

$$|\psi\rangle = \sum_{i_1,\ldots,i_n} P_n^{i_n} \ldots P_1^{i_1} |\psi\rangle \,, \tag{21}$$

which are guaranteed to be orthogonal by the consistency condition 3. We can thus find at least one set of orthogonal projection operators $\{R_I\}$, indexed by sets of the form $I = \{i_1 \ldots i_n\}$, which project onto the history vectors and sum to the identity:

$$\begin{aligned} R_{i_1 \ldots i_n} |\psi\rangle &= P_n^{i_n} \ldots P_1^{i_1} |\psi\rangle \,, \\ \sum_I R_I &= I \,, \\ R_I R_J &= \delta_{IJ} R_J \,. \end{aligned} \tag{22}$$

We can thus [32] construct a set $\mathcal{S}'$, with the same history vectors and the same probabilities as $\mathcal{S}$, built from a nested sequence of commuting projections defined by sums of the R_I. And, as we have seen, sets of this type are ordered consistent.

There is no reason to expect the projections defining the set $\mathcal{S}'$ to be closely related to those defining $\mathcal{S}$. In particular, the fact that $\mathcal{S}$ is a quasiclassical domain certainly does not imply that $\mathcal{S}'$ is likely to be: its projections are not generally likely to be interpretable in terms of familiar variables. But, as we have already noted, we have no theoretical criterion which identifies a particular consistent set, or a particular type of variable, as fundamentally correct for representing the events we observe. The set $\mathcal{S}'$ correctly identifies the history vectors and predicts their probabilities, and we thus could not say for certain (given that we presently have no theory of set selection) that its description of physics is fundamentally incorrect, while that given by $\mathcal{S}$ is fundamentally correct.

To make this observation is merely to point out a logical possibility. In fact, it would be extremely puzzling if the more complicated and apparently derivative set $\mathcal{S}'$ were in some sense more fundamental than the associated quasiclassical domain $\mathcal{S}$. And even if this were somehow understood to be true in principle, we would still need to understand the relationship between between ordered consistency and quasiclassicality in order to say whether or not standardly used subspace inferences are in fact justifiable.

8 Ordering and Ordering Violations: Interpretation

However, one of the main points of this paper — and the main reason for taking a particular interest in the properties of ordered consistent sets — is that either answer leads to an interesting conclusion.

If our empirical observations can be accounted for by the predictions of an ordered consistent set, then the ordered consistent sets formalism supersedes the current versions of the consistent histories formalism as a predictive theory. Alternatively, if the predictions of ordered consistent histories quantum theory are false, then either the consistent histories framework uses entirely the wrong language to describe histories of events in quantum theory, or we cannot generally rely on subspace implications in analysing our observations.

Either of these last two possibilities would have far-reaching implications for our understanding of nature. It is true that other ways of representing quantum histories are known than those used in the consistent histories formalism, but they arise either in non-standard versions of quantum theory, such as de Broglie–Bohm theory, or in alternative theories. It is also true that there is no way of logically excluding the possibility that subspace implications generally fail to hold. Any clear violation would, however, lead to radical changes in our representation of the world, and in particular to our understanding of the relation between theory and empirical observation.

I would suggest, however, that any version of closed system quantum theory in which the two orderings disagree leads to radical new interpretational problems. The fundamental problem is that, supposing that the world we experience is described by one particular realised quantum history, we never know — no matter how precise we try to make our observations — exactly what form that history takes. This is not only because we can never completely eliminate imprecision from our experimental observations. A deeper problem is that we have no theoretical understanding of how, precisely, an observation should be represented within quantum theory. We do not know precisely when and where any given observation takes place. Nor do we know whether is fundamentally correct to represent quantum events by projection operators, by quantum effects, by statements associated to space-time regions in path integral quantum theory, or in some other way — let alone precisely which operator, effect, or statement correctly represents any given event. As a result, we are always forced into guesswork and approximation. We are forced to assume, at least as a working hypothesis, that we can find sensible bounds on our observations. Roughly speaking, we assume that we can say, at least, that a photon hit our photographic plate within a certain region, that the observed flux from a distant star was in a certain range, and so forth.[8] We assume also that the probability of the actual observations — whose precise form we do not know — is bounded by the probability of the observations as we approximately represent them. These assumptions ultimately rely on the agreement of two orderings just mentioned: when those orderings disagree, we therefore run into new problems.

It is easy to see, in particular, that this sort of problem arises in any careful consistent histories treatment of quantum cosmology. Suppose, for example, that we have a sequence of cosmological events which we wish to represent theoretically, in order to calculate their probability, given some theory of the boundary conditions. Assuming that the basic principles of the consistent histories formalism are correct, we know that these events should be represented by some history H belonging to some consistent set $\mathcal{S}$. We do not, however, know the precise form of H or of $\mathcal{S}$: the events are given to us as empirical observations rather than as mathematical constructs.

The best we can then do, following the general principles of the consistent histories formalism, is to choose some plausible consistent set $\mathcal{S}'$ containing histories $H_{\rm min}$ and $H_{\rm max}$ which we guess to have the property $H_{\rm min} < H < H_{\rm max}$: in particular, thus, we choose $H_{\rm min} < H_{\rm max}$. Since $H_{\rm min}$ and $H_{\rm max}$ belong to the same set, we have that $p(H_{\rm min}) < p(H_{\rm max})$. It might naively be hoped that we can derive that $p(H_{\rm min}) < p(H) < p(H_{\rm max})$, but since H in general will belong to a different consistent set from $H_{\rm min}$ and $H_{\rm max}$, this does not generally follow. There is no way to bound $p(H)$, except (in principle) by performing the enormous task of explicitly calculating the probabilities of *all* consistent histories bounded by $H_{\rm min}$ and $H_{\rm max}$, and there is no way to justify the type of subspace implication — relating observations and true data — that we generally take for granted.

This is not to say that the disagreement of the two orderings necessarily leads to logical contradiction. Versions of quantum theory in which the orderings disagree need not be inconsistent, or even impossible to test precisely. They do, though, generally seem to require us to identify *precisely* the correct representation of our observations in quantum theory. This is generally a far from trivial problem: how are we to tell, a priori, exactly which projection operators represent the results of a series of quantum measurements? It is not impossible to imagine that theoretical criteria could be found which solve the problem, but we certainly do not have such criteria at present.

[8] In fact such statements are generally made within statistical confidence limits. To consider statistical statements would complicate the discussion a little, but does not alter the underlying point.

9 Conclusions

Though the criterion of ordering seems mathematically natural, both in the consistent histories approach to quantum theory and in other possible treatments of quantum histories, it raises very unconventional questions. It seems, though, that these questions cannot be avoided in any precise formulation of the quantum theory of a closed system which involves a standard representation of quantum events and which gives a historical account.

There seem to be three possibilities, each of which is interesting. The first is that the representations of quantum histories discussed here, though standard, are not those chosen by nature. Clearly this is a possibility: there are, for example, well known non-standard versions of quantum theory [6], and related theories [7], in which histories are defined by trajectories or other auxiliary variables, and in which subspace implications follow just as in classical physics.

The second possibility is that our quasiclassical domain can be shown to be an ordered consistent set. If so, then the ordered consistent histories approach is both predictively stronger than the standard consistent histories approach — since there are fewer ordered consistent sets — and compatible with empirical observation, and hence superior as a predictive theory. If it is compatible with our observations, the ordered consistent histories approach would seem at least as natural as the consistent histories approach.

Even if so, I would not suggest that the ordered consistent histories formalism is the "right" interpretation of quantum theory, and the consistent histories approach the "wrong" one. The ordered consistent histories approach seems almost certain to suffer from many of the same defects as the consistent histories approach, since there are still far too many ordered consistent sets. The aim here is thus not to propose the ordered consistent histories approach as a plausible fundamental interpretation of quantum theory, but to suggest that the range of natural and useful mathematical definitions of types of quantum history is wider than previously understood. This range includes, at least, Goldstein and Page's criterion of linear positivity [34], the various consistency criteria [1,26,32] in the literature, and the criterion of ordered consistency introduced here. It seems to me hard to justify taking any of these criteria as defining the fundamentally correct interpretation of quantum theory. On the one hand, physically interesting quantum histories might possibly satisfy any one, or none, of them; on the other hand, most quantum histories satisfying any given criterion seem unlikely to be physically interesting — and precisely which criteria are useful in which circumstances largely remains to be understood.

The third possibility is that our quasiclassical domain is not an ordered consistent set. This would have intriguing theoretical implications. We would have, at least in principle, to abandon subspace inferences, and we would ultimately need to understand precisely how to characterise the quantum events which constitute the history we observe. This would raise profound and not easily answerable questions about how we can tell what, precisely, *are* our empirical observations. It might also, depending on the way in which ordered consistency was violated, and the extent of any violation, raise significant practical problems in the analysis of those observations.

No compelling argument in favour of any one of these possibilities has been given here: it has been shown only that, if quasiclassical sets generally fail to be ordered consistent, they do so in a way too subtle to be displayed in the simplest models.

Another caveat is that the above discussion applied the criterion of ordered consistency only to the simplest representation of quantum histories, in which individual events are represented by projections at a single time. Other representations need to be considered case by case, and our conclusions might not necessarily generalize. For example, the fact that a consistent history built from single time projections is ordered when compared to consistent histories of the same type does not necessarily imply that it is ordered when compared to consistent histories defined by composite events.

Still, the criterion of ordered consistency defines a new version of the consistent histories formulation of quantum theory, which avoids the problems caused by contrary inferences. Its other properties and implications largely remain to be understood.

Acknowledgements

I am grateful to Jeremy Butterfield for a critical reading of the manuscript and many thoughtful comments, to Chris Isham and Noah Linden for very helpful discussions of their related work, and to Fay Dowker, Arthur Fine and Bob Griffiths for helpful comments.

I would particularly like to thank Francesco Petruccione for organising the small and lively meeting at which this work, inter alia, was discussed, and for his patient editorial encouragement.

This work was supported by a Royal Society University Research Fellowship.

Appendix: Ordering and Decoherence Functionals

This Appendix describes a noteworthy earlier discussion of quantum history orderings, given by Isham and Linden in Sec. IV of Ref. [14], and its relation to the ideas discussed here.

Isham and Linden abstract the basic ideas of the consistent histories formalism in the following way. First, the space $\mathcal{UP}$ of *history propositions* is taken to be a mathematical structure — an orthoalgebra — with a series of operations and relations obeying certain axioms. In particular, they propose that a partial ordering $\leq$ and an orthogonality relation $\perp$ should be defined on $\mathcal{UP}$ and should obey natural rules, and that $\mathcal{UP}$ should include an identity history 1.

They then introduce a space $\mathcal{D}$ of decoherence functionals, defined to be maps from $\mathcal{UP} \times \mathcal{UP}$ to the complex numbers satisfying certain axioms, and go on to consider whether the axioms defining decoherence functionals should include axioms relating to the ordering in $\mathcal{UP}$.

In the language of standard quantum mechanics, $\mathcal{UP}$ corresponds to the space of all the quantum histories (not only the consistent histories) for a given system, whose Hilbert space and hamiltonian are fixed. Any of several representations of quantum histories could be considered: the relevant part of Isham and Linden's discussion uses the simplest representation of quantum histories, as sequences of projection operators.

The standard quantum mechanical decoherence functional (as appears on the left hand side of (3)) is a member of the space $\mathcal{D}$ in the minimal axiom system Isham and Linden eventually choose. As they remark, though, it would not be a member of $\mathcal{D}$ if the extra ordering axioms they discuss were imposed. Isham and Linden nonetheless consider imposing these ordering axioms, since their aim in the relevant discussion is to investigate generalised algebraic and logical schemes rather than to propose a formalism applicable to standard quantum theory. (They suggest, at the end of section IV, that standard quantum theory might perhaps emerge from some such generalised scheme in an appropriate limit.)

Isham and Linden were, as far as I am aware, the first to investigate possible uses of orderings in developing the consistent histories formalism. It is worth stressing, though, to avoid any possible confusion, that their suggestions pursue the exploration of orderings in a direction orthogonal to the one considered in the present paper. In this paper we restrict attention to standard quantum theory, and propose an alternative histories formalism within that theory, using the standard quantum theoretic decoherence functional throughout. We note also that the subspace implications which underlie our basic scientific worldviews depend for their justification on the assumption that the quasiclassical set describing the physics we observe is an ordered consistent set. Isham and Linden's proposed ordering axioms, on the other hand, exclude standard quantum theory and the standard decoherence functional: they are possible postulates which might be imposed on non-standard generalised decoherence functionals in non-standard generalisations of quantum theory.

Isham and Linden give a minimal set of postulated properties for generalised decoherence functionals:

$$\begin{aligned}
&d(0,\alpha) = 0 \text{ for all } \alpha \in \mathcal{UP}\,;\\
&d(\alpha,\beta) = d(\beta,\alpha)^* \text{ for all } \alpha,\beta \in \mathcal{UP}\,;\\
&d(\alpha,\alpha) \geq 0 \text{ for all } \alpha \in \mathcal{UP}\,;\\
&\text{if } \alpha \perp \beta \text{ then, for all } \gamma,\ d(\alpha \oplus \beta,\gamma) = d(\alpha,\gamma) + d(\beta,\gamma)\,;\\
&d(1,1) = 1\,.
\end{aligned} \qquad (23)$$

They then consider imposing new postulates on decoherence functionals. The first of these — their posited inequality 1 — is that:

$$\text{for all } d \in \mathcal{D} \text{ and for all } \alpha, \beta \text{ with } \alpha \leq \beta \text{ we have } d(\alpha,\alpha) \leq d(\beta,\beta)\,. \quad (24)$$

As Isham and Linden go on to point out, there are familiar examples in standard quantum theory in which (24) is violated for a pair of histories $\alpha \leq \beta$ in which one of the histories (in their case α) is inconsistent.[9]

Two further postulates on generalised decoherence functionals are also posited:

$$\alpha \perp \beta \text{ implies } d(\alpha,\alpha) + d(\beta,\beta) \leq 1 \text{ for all } d \in \mathcal{D}\,; \quad (25)$$

and

$$\text{for all } d \in \mathcal{D} \text{ and all } \gamma \in \mathcal{UP} \text{ we have } d(\gamma,\gamma) \leq 1\,. \quad (26)$$

Isham and Linden give examples to show that, in standard quantum theory, with the standard decoherence functional, inconsistent histories do not necessarily respect these inequalities either.

Again, the difference from the examples considered in the present paper is worth emphasizing. All of the examples Isham and Linden consider involve inconsistent histories — these are all they require in order to investigate possible properties of decoherence functionals applied to arbitrary, not necessarily consistent, quantum histories. These examples are not problematic for the consistent histories approach to quantum theory, according to which the inconsistent histories have no physical significance, and they do not give rise to new interpretational questions in any conventional quantum histories approach, for essentially the same reason. The discussion in the present paper, on the other hand, looks at the properties of *consistent* histories in standard quantum theory: we have argued that their failure to respect ordering relations is problematic and explained that it does raise new questions.

Suppose now that we set aside Isham and Linden's motivations, and alter their ordering postulates so that they apply, not to generalised decoherence functionals applied to all quantum histories in an abstract generalisation of quantum theory, but to the standard decoherence functional applied to ordered consistent histories in standard quantum theory. We then obtain the following:

$$\text{for all } \alpha, \beta \text{ with } \alpha \leq \beta \text{ we have } d(\alpha,\alpha) \leq d(\beta,\beta)\,; \quad (27)$$

$$\alpha \perp \beta \text{ implies } d(\alpha,\alpha) + d(\beta,\beta) \leq 1\,; \quad (28)$$

and

$$\text{for all } \gamma \text{ we have } d(\gamma,\gamma) \leq 1\,. \quad (29)$$

[9] The suggestion that inequality 1 is true when applied to sequences of projectors onto subsets of configuration space in a path-integral quantum theory is thus misleading: it is easy to find configuration space analogues of these examples. I am grateful to Chris Isham and Noah Linden for discussions of this point.

Here d is the standard decoherence functional, α, β and γ are now taken to be ordered consistent histories, and $\alpha \perp \beta$ means that α and β are disjoint — i.e., there is at least one time at which their respective events are represented by contrary projections.

The first of these equations holds by the definition of an ordered consistent history, but it might perhaps be hoped that the others could restrict the class of histories further. However, the second equation also holds for all ordered consistent histories. To see this, note that $\alpha \perp \beta$ implies that $\beta \leq (1-\alpha)$, and that if α is a consistent history then $(1-\alpha)$ is too. The fact that β is ordered consistent thus implies that

$$p(\beta) \leq p(1-\alpha) = 1 - p(\alpha)\,. \tag{30}$$

The third equation, moreover, holds for all consistent histories, ordered or otherwise. It seems, then, that ordered consistency may be the strongest natural criterion that can be defined using the basic ingredients of consistency and ordering.

References

1. Griffiths R. B. (1984): J. Stat. Phys. **36**, 219.
2. Griffiths R. B. (1993): Found. Phys. **23**, 1601.
3. Omnès R. (1988): J. Stat. Phys. **53**, 893.
4. Gell-Mann M., Hartle J. B. (1990): In: Zurek W. (Ed.). *Complexity, Entropy, and the Physics of Information, SFI Studies in the Sciences of Complexity*, Vol. VIII, Addison Wesley, Reading.
5. Isham C. J. (1994): J. Math. Phys. **23**, 2157.
6. Bohm D. (1952): Phys. Rev. **85**, 166.
7. Ghirardi G. , Rimini A. and Weber T. (1986): Phys. Rev. **D34**, 470.
8. Sorkin R. D. (1994): Mod. Phys. Lett. **A9**, 3119.
9. Dowker F., Kent A. (1996): J. Stat. Phys. **82**, 1575.
10. Kent A. (1998): In: *Modern Studies of Basic Quantum Concepts and Phenomena*, Proceedings of the 104th Nobel Symposium, Gimo, June 1997, Physica Scripta **T76**, 78-84.
11. Rudolph O. (1996): Int. J. Theor. Phys. **35**, 1581.
12. Rudolph O. (1996): J. Math. Phys. **37**, 5368.
13. Hartle J. B. (1991): In: Coleman S., Hartle J., Piran T., Weinberg S. (Eds.), *Quantum Cosmology and Baby Universes*, Proceedings of the 1989 Jerusalem Winter School on Theoretical Physics, World Scientific, Singapore.
14. Isham C. J., Linden N. (1994): J. Math. Phys. **35**, 5452.
15. Isham C. J., Linden N., Schreckenberg S. (1994): J. Math. Phys. **35**, 6360.
16. Omnès R. (1994): *The Interpretation of Quantum Mechanics*, Princeton University Press, Princeton.
17. Griffiths R. B. (1996): Phys. Rev. **A54**, 2759.
18. Kent A. (1996): Phys. Rev. **A54**, 4670.
19. Kent A. (1997): Phys. Rev. Lett. **78**, 2874.
20. Dowker F., Kent A. (1995): Phys. Rev. Lett. **75**, 3038.

21. Paz J., Zurek W. (1993): Phys. Rev. **D48**, 2728.
22. Giardina I., Rimini A. (1996): Found. Phys. **26**, 973.
23. Kent A. (1996): In: Cushing J., Fine A., Goldstein S. (Eds.), *Bohmian Mechanics and Quantum Theory: An Appraisal*, Kluwer Academic Press, Dordrecht.
24. Omnès R. (1994): Phys. Lett. **A 187**, 26.
25. Kent A., McElwaine J. (1997): Phys. Rev. **A55**, 1703.
26. Gell-Mann M., Hartle J. B. (1995): University of California, Santa Barbara preprint UCSBTH-95-28; gr-qc/9509054.
27. Griffiths R., Hartle J. (1998): Phys. Rev. Lett. **81**, 1981.
28. Kent A. (1998): Phys. Rev. Lett. **81**, 1982.
29. Cohen O. (1995): Phys. Rev. **A51**, 4373.
30. Aharonov Y., Vaidman L. (1991): J. Phys. **A 24**), 2315.
31. Gell-Mann M., Hartle J. B. (1994): University of California, Santa Barbara preprint UCSBTH-94-09; gr-qc/9404013.
32. Gell-Mann M., Hartle J. B. (1993): Phys. Rev. **D47**, 3345.
33. McElwaine J. (1997): Phys. Rev. **A56**, 1756.
34. Goldstein S., Page D. (1995): Phys. Rev. Lett. **74**, 3715.

Quantum Measurements and Non-locality

Sandu Popescu[1,2] and Nicolas Gisin[3]

[1] H.H. Wills Physics Laboratory, University of Bristol, Tyndall Avenue, Bristol BS8 1TL, UK
[2] BRIMS, Hewlett-Packard Laboratories, Stoke Gifford, Bristol BS12 6QZ, UK
[3] Group of Applied Physics, University of Geneva, 1211 Geneva 4, Switzerland

Abstract. We discuss the role of non-locality in the problem of determining the state of a quantum system, one of the most basic problems in quantum mechanics.

1 Introduction

In 1964 J. Bell [1] introduced the idea of non-local correlations. He considered the situation in which measurements are performed on pairs of particles which were prepared in entangled states and the two members of the pair are separated in space, and he showed that the results of the measurement are correlated in a way which cannot be explained by any local models. Since then, non-locality has been recognized as one of the most important aspects of quantum mechanics.

Following Bell, most studies of non-locality focused on situations when two or more particles are separated in space and are prepared in entangled states. Entanglement appeared to be a sine-qua-non requirement for non-locality. Indeed, all entangled states were shown to be non-local [2] while all direct product states lead to purely local correlations.

Surprisingly enough, however, it turns out that non-locality plays an essential role in problems in which there is no entanglement whatsoever and apparently everything is local. One such problem - probably one of the most basic problems in quantum mechanics - is that of determining by measurements the state of a quantum system. That is, given a quantum system in some state Ψ unknown to us, how can we determine what Ψ is? As is well-known, because quantum measurements yield probabilistic outcomes, and due to the fact that measuring an observable we disturb all the other non-commuting observables, we cannot determine with certainty the unknown state Ψ out of measurements on a single particle. If we have a large number of identically prepared particles - a so called "quantum ensemble" - by performing measurements on the different particles we can accumulate enough results so that from their statistics we could determine Ψ. But anything less than an infinite number of particles limits our ability to determine the state with certainty.

Now, although when we have only a finite ensemble of identically prepared particles we cannot determine their state with absolute precision, the question remains of how well can we determine the state and which is the the measurement we should perform. It is in this context that, unexpectedly, non-locality creeps in.

The story of interest here begins in 1991 when Peres and Wootters [3] considered the following particular question of quantum state estimation. Suppose that we have a quantum ensemble which consists of only two identically prepared spin 1/2 particles. In other words, we are given two spin 1/2 particles, and we are told that the spins are parallel, but we are not told the polarization direction. What is the best way to find the direction? One possibility would involve measurements carried out on each spin separately, and trying to infer the polarization direction based on the results of the two measurements. Another possibility would be again to measure the two spins separately, but to measure first one spin and then choose some appropriate measurement on the second spin depending on the result of the measurement on the first spin. Finally, they conjectured, an even better way might be to measure the two spins together, not separately. That is, to perform a measurement of an operator whose eigenstates are *entangled* states of the two spins. The Peres-Wootters conjecture was subsequently proven by Massar and Popescu [4].

What we see here is a surprising manifestation of non-locality. Indeed, suppose that the two spins are separate in space. They are prepared in *direct-product* states (they are each polarized along the same direction). Nevertheless, by simply performing local measurements one cannot extract all information from the spins. The optimal measurement *must* introduce non-locality by projecting them onto entangled states.

The above effect has opened a new direction of research, and its implications are just starting to be uncovered. In this paper we will discuss a new surprising effect [5] whose existence is made possible by the fact that optimal measurements imply non-locality.

2 Measurements on 2-Particle Systems with Parallel or Anti-Parallel Spins

The problem we consider here is the following. Suppose Alice wants to communicate Bob a space direction $\boldsymbol{n}$. She may do that in two ways. In the first case, Alice sends Bob two spin 1/2 particles polarized along $\boldsymbol{n}$, i.e. $|\boldsymbol{n}, \boldsymbol{n}\rangle$. When Bob receives the spins, he performs some measurement on them and then guesses a direction $\boldsymbol{n}_g$ which has to be as close as possible to the true direction $\boldsymbol{n}$. The second method is almost identical to the first, with the difference that Alice sends $|\boldsymbol{n}, -\boldsymbol{n}\rangle$, i.e. the first spin is polarized along $\boldsymbol{n}$ but the second one is polarized in the *opposite* direction. The question is whether these two methods are equally good or, if not, which is better[1].

[1] It might be the case that there is some other method which is better for transmitting directions than the two ones mentioned above. However, we are not

For a better perspective, consider first a simpler problem. Suppose Alice wants to communicate Bob a space direction $\boldsymbol{n}$ and she may do that by one of the following two strategies. In the first case, Alice sends Bob a single spin 1/2 particle polarized along $\boldsymbol{n}$, i.e. $|\boldsymbol{n}\rangle$. The second strategy is identical to the first, with the difference that when Alice wants to communicate Bob the direction $\boldsymbol{n}$ she sends him a single spin 1/2 particle polarized in the opposite direction, i.e. $|-\boldsymbol{n}\rangle$. Which of these two strategies is better?

Obviously, if the particles would be classical spins then, both methods would be equally good, as an arrow defines equally well both the direction in which it points and the opposite direction. Is the quantum situation the same?

First, we should note that in general, by sending a single spin 1/2 particle, Alice cannot communicate to Bob the direction $\boldsymbol{n}$ with absolute precision. Nevertheless, it is still obviously true that the two strategies are equally good. Indeed, all Bob has to do in the second case is to perform exactly the same measurements as he would do in the first case, only that when his results are such that in the first case he would guess $\boldsymbol{n}_g$, in the second case he guesses $-\boldsymbol{n}_g$.

One is thus tempted to think that, similar to the classical case, for the purpose of defining a direction $\boldsymbol{n}$, a quantum mechanical spin polarized along $\boldsymbol{n}$ is as good as a spin polarized in the opposite direction: in particular, the two two-spin states $|\boldsymbol{n},\boldsymbol{n}\rangle$ and $|\boldsymbol{n},-\boldsymbol{n}\rangle$ should be equally good. Surprisingly however this is not true.

That there could be any difference between communicating a direction by two parallel spins or two anti-parallel spins seems, at first sight, extremely surprising. After all, by simply flipping one of the spins we could change one case into the other. For example, if Bob knows that Alice indicates the direction by two anti-parallel spins he only has to flip the second spin and then apply all the measurements as in the case in which Alice sends from the beginning parallel spins. Thus, apparently, the two methods are bound to be equally good.

The problem is that one cannot flip a spin of unknown polarization. Indeed, the flip operator V defined as

$$V|\boldsymbol{n}\rangle = |-\boldsymbol{n}\rangle \tag{1}$$

is not unitary but anti-unitary. To see this we note that in Heisenberg representation flipping the spin means changing the sign of all spin operators, $\boldsymbol{\sigma} \to -\boldsymbol{\sigma}$. But this transformation does not preserve the commutation relations, i.e.

$$[\sigma_x, \sigma_y] = \mathrm{i}\sigma_z \to [\sigma_x, \sigma_y] = -\mathrm{i}\sigma_z, \tag{2}$$

---

interested here in finding an optimal method for communicating directions; we are only interested in comparing the parallel and anti-parallel spins methods.

while all unitary transformations leave the commutation relations unchanged. Thus there is no physical operation which could implement such a transformation.

A couple of questions arise. First, why is it still the case that a single spin polarized along $\boldsymbol{n}$ defines the direction as well as a single spin polarized in the opposite direction? The reason is that although Bob cannot implement an *active* transformation, i.e. cannot flip the spin, he can implement a *passive* transformation: he can flip his measuring devices. Indeed, there is no problem for Bob in flipping all his Stern-Gerlach apparatuses, or, even simpler than that, to merely rename the outputs of each Stern-Gerlach "up"→ "down" and "down"→"up".

At this point it is natural to ask why can't Bob solve the problem of two spins in the same way, namely by performing a passive transformation on the apparatuses used to measure the second spin? The problem is the entanglement. Indeed, if the optimal strategy for finding the polarization direction would involve separate measurements on the two spins then two parallel spins would be equivalent to two anti-parallel spins. (This would be true even if which measurement is to be performed on the second spin depends on the result of the measurement on the first spin.) But, as explained in the Introduction, the optimal measurement is *not* a measurement performed separately on the two spins but a measurement which acts on both spins simultaneously, that is, the measurement of an operator whose eigenstates are entangled states of the two spins. For such a measurement there is no way of associating different parts of the measuring device with the different spins, and thus there is no way to make a passive flip associated to the second spin. Consequently there is no way, neither active nor passive to implement an equivalence between the parallel and anti-parallel spin cases.

After understanding that there is indeed room for the two direction communication methods to be different, let us now investigate them in detail.

To start with, we have to define some figure of merit which will tell us how successful a communication protocol is. For concreteness, let us define Bob's measure of success as the fidelity

$$F = \int \mathrm{d}\boldsymbol{n} \sum_g P(g|\boldsymbol{n}) \frac{1 + \boldsymbol{n}\boldsymbol{n}_g}{2} \tag{3}$$

where $\boldsymbol{n}\boldsymbol{n}_g$ is the scalar product in between the true and the guessed directions, the integral is over the different directions $\boldsymbol{n}$ and $\mathrm{d}\boldsymbol{n}$ represents the *a priori* probability that a state associated to the direction $\boldsymbol{n}$, i.e. $|\boldsymbol{n},\boldsymbol{n}\rangle$ or $|\boldsymbol{n},-\boldsymbol{n}\rangle$ respectively, is emitted by the source; $P(g|\boldsymbol{n})$ is the probability of guessing $\boldsymbol{n}_g$ when the true direction is $\boldsymbol{n}$. In other words, for each trial Bob gets a score which is a (linear) function of the scalar product between the true and the guessed direction, and the final score is the average of the individual scores.

When the different directions $\boldsymbol{n}$ are randomly and uniformly distributed over the unit sphere, an optimal measurement for pairs of parallel spins $\psi = |\boldsymbol{n},\boldsymbol{n}\rangle$ has been found by Massar and Popescu [4]. Bob has to measure an operator A whose eigenvectors ϕ_j, $j = 1,\ldots,4$, are

$$|\phi_j\rangle = \frac{\sqrt{3}}{2}|\boldsymbol{n}_j,\boldsymbol{n}_j\rangle + \frac{1}{2}|\psi^-\rangle \tag{4}$$

where $|\psi^-\rangle$ denotes the singlet state and the Bloch vectors $\boldsymbol{n}_j$ point to the 4 vertices of the tetrahedron:

$$\begin{aligned} \boldsymbol{n}_1 &= (0,0,1), & \boldsymbol{n}_2 &= (\frac{\sqrt{8}}{3},0,-\frac{1}{3}), \\ \boldsymbol{n}_3 &= (\frac{-\sqrt{2}}{3},\sqrt{\frac{2}{3}},-\frac{1}{3}), & \boldsymbol{n}_4 &= (\frac{-\sqrt{2}}{3},-\sqrt{\frac{2}{3}},-\frac{1}{3}). \end{aligned} \tag{5}$$

The phases used in the definition of $|\boldsymbol{n}_j\rangle$ are such that the 4 states ϕ_j are mutually orthogonal. The exact values of the eigenvalues corresponding to the above eigenvectors are irrelevant; all that is important is that they are different from each other, so that each eigenvector can be unambiguously associated to a different outcome of the measurement. If the measurement results corresponds to ϕ_j, then the guessed direction is $\boldsymbol{n}_j$. The corresponding optimal fidelity is 3/4 [4].

A related case is when the directions $\boldsymbol{n}$ are *a priori* on the vertices of the tetrahedron, with equal probability 1/4. Then the above measurement provides a fidelity of $5/6 \approx 0.833$, conjectured to be optimal.

Let us now consider pairs of anti-parallel spins, $|\psi >= |\boldsymbol{n},-\boldsymbol{n}\rangle$, and the measurement whose eigenstates are

$$\theta_j = \alpha|\boldsymbol{n}_j,-\boldsymbol{n}_j\rangle - \beta\sum_{k\neq j}|\boldsymbol{n}_k,-\boldsymbol{n}_k\rangle \tag{6}$$

with $\alpha = \frac{13}{6\sqrt{6}-2\sqrt{2}} \approx 1.095$ and $\beta = \frac{5-2\sqrt{3}}{6\sqrt{6}-2\sqrt{2}} \approx 0.129$. The corresponding fidelity for uniformly distributed $\boldsymbol{n}$ is $F = \frac{5\sqrt{3}+33}{3(3\sqrt{3}-1)^2} \approx 0.789$; and for $\boldsymbol{n}$ lying on the tetrahedron $F = \frac{2\sqrt{3}+47}{3(3\sqrt{3}-1)^2} \approx 0.955$. In both cases the fidelity obtained for pairs of anti-parallel spins is larger than for pairs of parallel spins!

It is useful now to investigate in more detail what is going on. We have claimed above that when we perform a measurement of an operator whose eigenstates are entangled states of the two spins, there is no way of making a passive flip associated with the second spin. We would like to comment more about this point.

It is clear that in the case of a measuring device corresponding to an operator whose eigenstates are entangled states of the two spins, we cannot identify one part of the apparatus as acting solely on one spin and another part of the apparatus as acting on the second spin. Thus we cannot simply isolate a part of the measuring device and rename its outcomes. But perhaps one could make such a passive transformation at a *mathematical* level, that is, in the mathematical description of the operator associated to the measurement and then physically construct an apparatus which corresponds to the new operator.

In the case of two parallel spins the optimal measurement is described by a nondegenerate operator whose eigenstates $|\phi_j\rangle$ are given by (4) and (5). It is convenient to consider the projectors $P^j = |\phi_j\rangle\langle\phi_j|$ associated with the eigenstates. As is well-known, any unit-trace hermitian operator, and in particular any projector, can be written as

$$P^j = \frac{1}{4}(I + \boldsymbol{\alpha}^j \boldsymbol{\sigma}^{(1)} + \boldsymbol{\beta}^j \boldsymbol{\sigma}^{(2)} + R^j_{k,l}\sigma_k^{(1)}\sigma_l^{(2)}). \tag{7}$$

with some appropriate coefficients $\boldsymbol{\alpha}^j$, $\boldsymbol{\beta}^j$ and $R^j_{k,l}$. (The upper indexes on the spin operators mean "particle 1" or "2" and I denotes the identity). Why then couldn't we simply make the passive spin flip by considering a measurement described by the projectors

$$\tilde{P}^j = \frac{1}{4}(I + \boldsymbol{\alpha}^j \boldsymbol{\sigma}^{(1)} - \boldsymbol{\beta}^j \boldsymbol{\sigma}^{(2)} - R^j_{k,l}\sigma_k^{(1)}\sigma_l^{(2)}). \tag{8}$$

obtained by the flip of the operators associated second spin, $\boldsymbol{\sigma}^{(2)} \to -\boldsymbol{\sigma}^{(2)}$? The reason is that the transformed operators $\tilde{P}^j$ are no longer projectors! Indeed, each projection operator P^j could also be viewed as a density matrix $\rho^j = P^j = |\phi_j\rangle\langle\phi_j|$. The passive spin flip which leads from Eq. (7) to Eq. (8) is nothing more than the partial transpose of the density matrices ρ^j with respect to the second spin. But each density matrix ρ^j is *non-separable* (because it describes the entangled state $|\phi_j\rangle$). However, according to the well-known result of the Horodeckis [6,7] the partial transpose of a non-separable density matrix of two spin 1/2 particles has a negative eigenvalue and thus it cannot represent a projector anymore. On the other hand, if the optimal measurement would have consisted of independent measurements on the two spins, each projector would have been a direct product density matrix and the spin flip would have transformed them into new projectors, and thus led to a valid new measurement.

The above analysis of encoding directions by parallel or anti-parallel spins shows a most important aspect of the problem of state estimation. Consider the two sets of states, that of parallel spins and that of anti-parallel spins. The distance in between any two states in the first set is equal to the distance in between the corresponding pair of states in the second set. That is,

$$|\langle \boldsymbol{n}, \boldsymbol{n}|\boldsymbol{m}, \boldsymbol{m}\rangle|^2 = |\langle \boldsymbol{n}, -\boldsymbol{n}|\boldsymbol{m}, -\boldsymbol{m}\rangle|^2. \tag{9}$$

Nevertheless, *as a whole*, the anti-parallel spin states are further apart than the parallel ones! Indeed, the anti-parallel spin states span the entire 4-dimensional Hilbert space of the two spin 1/2, while the parallel spin states span only the 3-dimensional subspace of symmetric states. This is similar to a 3-spin example discovered by R. Jozsa and J. Schlienz [8].

Furthermore, suppose we consider a simpler problem in which Alice has to communicate Bob one out of only two possible directions, $\boldsymbol{n}$ and $\boldsymbol{m}$. Then, since $|\langle \boldsymbol{n}, \boldsymbol{n}|\boldsymbol{m}, \boldsymbol{m}\rangle|^2 = |\langle \boldsymbol{n}, -\boldsymbol{n}|\boldsymbol{m}, -\boldsymbol{m}\rangle|^2$ the parallel and anti-parallel spins methods would be equally good. The two methods are also equally good in the case when Alice has to communicate to Bob an arbitrary direction in a plane. Indeed, suppose that the directions Alice has to communicate are restricted to the $x-y$ plane. Then a rotation of the second spin by 180 degrees around the z axis can transform any state of parallel spins into a state of anti-parallel ones and vice-versa. It is only when Alice has to communicate directions which do not lie in the same plane, that the two methods become different. This shows that the problem of state estimation depends on the *global* structure of the set of states under investigation and cannot be reduced to the problem of pairwise distinguishibility.[2]

3 Conclusions

To conclude, we have shown that non-locality plays a fundamental role in what is probably the most basic quantum mechanical problem - determining the state of a quantum system. The link between non-locality and measurement theory is completely unexpected, and without any doubt, it will lead to new insights into the very nature of quantum mechanics. Furthermore, it is already leading to possible practical applications in quantum communication [9].

Acknowledgments

SP would like to thank very warmly the wonderful hospitality offered by the Istituto Italiano per gli Studi Filosofici, Napoli.

References

1. Bell J. S. (1964): Physics **1**, 195.
2. Gisin N. (1991): Phys. Lett.A **154**, 201; Popescu S., Rohrlich D. (1992): Phys. Lett. A **166**, 293.
3. Peres A., Wootters W. (1991): Phys. Rev. Lett. **66**, 1119.
4. Massar S., Popescu S. (1995): Phys. Rev. Lett. **74**, 1259.
5. Gisin N., Popescu S. (1999): Phys. Rev. Lett. **83**, 432.
6. Peres A. (1996): Phys. Rev. Lett. **76**, 1413.
7. Horodecki M. R., Horodecki P. (1996): Phys. Lett. A **223**, 1.
8. Jozsa R., Schlienz J. (1999): quant-ph/9911009.
9. Rudolph T. (1999): quant-ph/9902010.

[2] Technically this follows from the fact that quantum mechanical states are vectors in a complex Hilbert space rather than in a Hilbert space with real coefficients.

False Loss of Coherence

William G. Unruh

CIAR Cosmolgy Program, Dept. of Physics
University of B.C.
Vancouver, Canada V6T 1Z1

Abstract. The loss of coherence of a quantum system coupled to a heat bath as expressed by the reduced density matrix is shown to lead to the mis-characterization of some systems as being incoherent when they are not. The spin boson problem and the harmonic oscillator with massive scalar field heat baths are given as examples of reduced incoherent density matrices which nevertheless still represent perfectly coherent systems.

1 Massive Field Heat Bath and a Two Level System

How does an environment affect the quantum nature of a system? The standard technique is to look at the reduced density matrix, in which one has traced out the environment variables. If this changes from a pure state to a mixed state (entropy $-\mathrm{Tr}\rho \ln \rho$ not equal to zero) one argues that the system has lost quantum coherence, and quantum interference effects are suppressed. However this criterion is too strong. There are couplings to the environment which are such that this reduced density matrix has a high entropy, while the system alone retains virtually all of its original quantum coherence in certain experiments.

The key idea is that the external environment can be different for different states of the system. There is a strong correlation between the system and the environment. As usual, such correlations lead to decoherence in the reduced density matrix. However, the environment in these cases is actually tied to the system, and is adiabatically dragged along by the system. Thus although the state of the environment is different for the two states, one can manipulate the system alone so as to cause these apparently incoherent states to interfere with each other. One simply causes a sufficiently slow change in the system so as to drag the environment variables into common states so the quantum interference of the system can again manifest itself.

An example is if one looks at an electron with its attached electromagnetic field. Consider the electron at two different positions. The static Coulomb field of the two charges differ, and thus the states of the electromagnetic field differ with the electron in the two positions. These differences can be sufficient to cause the reduced electron wave function loose coherence for a state which is a coherent sum of states located at these two positions. However, if one causes the system to evolve so as to cause the electron in those two positions to come together (e.g., by having a force field such that the electron in both positions to be brought together at some central point for example), those two apparently incoherent states will interfere, demonstrating that the loss of coherence was not real.

Another example is light propagating through a slab of glass. If one simply looks at the electromagnetic field, and traces out over the states of the atoms in the glass, the light beams traveling through two separate regions of the glass will clearly decohere– the reduced density matrix for the electromagnetic field will lose coherence in position space– but those two beams of light will also clearly interfere when they exit the glass or even when they are within the glass.

The above is not to be taken as proof, but as a motivation for the further investigation of the problem. The primary example I will take will be of a spin-$\frac{1}{2}$ particle (or other two-level system). I will also examine a harmonic oscillator as the system of interest. In both cases, the heat bath will be a massive one dimensional scalar field. This heat bath is of the general Caldeira-Leggett type [1] (and in fact is entirely equivalent to that model in general). The mass of the scalar field will be taken to be larger than the inverse time scale of the dynamical behaviour of the system. This is not to be taken as an attempt to model some real heat bath, but to display the phenomenon in its clearest form. Realistic heat baths will in general also have low frequency excitations which will introduce other phenomena like damping and genuine loss of coherence into the problem.

2 Spin-$\frac{1}{2}$ System

Let us take as our first example that of a spin-$\frac{1}{2}$ system coupled to an external environment. We will take this external environment to be a one-dimensional massive scalar field. The coupling to the spin system will be via purely the 3-component of the spin. I will use the velocity coupling which I have used elsewhere as a simple example of an environment (which for a massless field is completely equivalent to the Caldeira-Leggett model). The Lagrangian is

$$L = \int \frac{1}{2} \left[(\dot\phi(x))^2 - \phi'(x)^2 - m^2\phi(x)^2 + 2\epsilon\dot\phi(x)h(x)\sigma_3 \right] \mathrm{d}x, \tag{1}$$

which gives the Hamiltonian

$$H = \int \frac{1}{2} \left[(\pi(x) - \epsilon h(x)\sigma_3)^2 + \phi'(x)^2 + m^2\phi(x)^2 \right] \mathrm{d}x. \tag{2}$$

$h(x)$ is the interaction range function, and its Fourier transform is related to the spectral response function of Leggett and Caldeira.

This system is easily solvable. I will look at this system in the following way. Start initially with the field in its free ($\epsilon = 0$) vacuum state, and the system is in the +1 eigenstate of σ_1. I will start with the coupling ϵ initially zero and gradually increase it to some large value. I will look at the reduced density matrix for the system, and show that it reduces to one which is almost the identity matrix (the maximally incoherent density matrix) for strong coupling. Now I let ϵ slowly drop to zero again. At the end of the procedure, the state of the system will again be found to be in the original eigenstate of σ_1. The intermediate maximally incoherent density matrix would seem to imply that the system no longer has any quantum coherence. However, this lack of coherence is illusionary. Slowly decoupling the system from the environment should in the usual course simply maintain the incoherence of the system. Yet here, as if by magic, an almost completely incoherent density matrix magically becomes coherent when the system is decoupled from the environment.

In analyzing the system, I will look at the states of the field corresponding to the two possible σ_3 eigenstates of the system. These two states of the field are almost orthogonal for strong coupling. However they correspond to fields tightly bound to the spin system. As the coupling is reduced, the two states of the field adiabatically come closer and closer together until finally they coincide when ϵ is again zero. The two states of the environment are now the same, there is no correlation between the environment and the system, and the system regains its coherence.

The density matrix for the spin system can always be written as

$$\rho(t) = \frac{1}{2}(1 + \boldsymbol{\rho}(t) \cdot \boldsymbol{\sigma}) \tag{3}$$

where

$$\boldsymbol{\rho}(t) = \mathrm{Tr}(\boldsymbol{\sigma}\rho(t)). \tag{4}$$

We have

$$\boldsymbol{\rho}(t) = \mathrm{Tr}\left(\boldsymbol{\sigma}\mathcal{T}\left[\mathrm{e}^{-\mathrm{i}\int_0^t H\mathrm{d}t}\right]\frac{1}{2}(1 + \boldsymbol{\rho}(0)\cdot\boldsymbol{\sigma})R_0\mathcal{T}\left[\mathrm{e}^{-\mathrm{i}\int H\mathrm{d}t}\right]^\dagger\right), \tag{5}$$

where R_0 is the initial density matrix for the field (assumed to be the vacuum), and $\mathcal{T}$ is the time-ordering operator. Because ϵ and thus H is time-dependent, the H's at different times do not commute. This leads to the requirement for the time-ordering in the expression. As usual, the time-ordered integral is the way of writing the time ordered product $\prod_n \mathrm{e}^{-\mathrm{i}H(t_n)\mathrm{d}t} = \mathrm{e}^{-\mathrm{i}H(t)\mathrm{d}t}\mathrm{e}^{-\mathrm{i}H(t-\mathrm{d}t)\mathrm{d}t}....\mathrm{e}^{-\mathrm{i}H(0)\mathrm{d}t}$.

Let us first calculate $\rho_3(t)$. We have

$$\begin{aligned}\rho_3(t) &= \mathrm{Tr}\left(\sigma_3\mathcal{T}\left[\mathrm{e}^{-\mathrm{i}\int_0^t H\mathrm{d}t}\right]\frac{1}{2}(1 + \boldsymbol{\rho}(0)\cdot\boldsymbol{\sigma})R_0\mathcal{T}\left[\mathrm{e}^{-\mathrm{i}\int H\mathrm{d}t}\right]^\dagger\right)\\ &= \mathrm{Tr}\left(\mathcal{T}[\mathrm{e}^{-\mathrm{i}\int_0^t H\mathrm{d}t}]\sigma_3\frac{1}{2}(1 + \boldsymbol{\rho}(0)\cdot\boldsymbol{\sigma})R_0\mathcal{T}[\mathrm{e}^{-\mathrm{i}\int H\mathrm{d}t}]^\dagger\right)\end{aligned}$$

$$= \mathrm{Tr}\left(\sigma_3 \frac{1}{2}(1+\boldsymbol{\rho}(0)\cdot\boldsymbol{\sigma})R_0\right)$$
$$= \rho_3(0) \tag{6}$$

because σ_3 commutes with $H(t)$ for all t. We now define

$$\sigma_+ = \frac{1}{2}(\sigma_1 + i\sigma_2) = |+\rangle\langle-|, \qquad \sigma_- = \sigma_+^\dagger. \tag{7}$$

Using $\sigma_+\sigma_3 = -\sigma_+$ and $\sigma_3\sigma_+ = \sigma_+$ we have

$$\mathrm{Tr}\left(\sigma_+ \mathcal{T}\left[\mathrm{e}^{-\mathrm{i}\int_0^t H\mathrm{d}t}\right]\frac{1}{2}(1+\boldsymbol{\rho}(0)\cdot\boldsymbol{\sigma})R_0\mathcal{T}\left[\mathrm{e}^{-\mathrm{i}\int H\mathrm{d}t}\right]^\dagger\right)$$
$$= \mathrm{Tr}_\phi\left(\mathcal{T}\left[\mathrm{e}^{-\mathrm{i}\int(H_0-\epsilon(t)\int\pi(x)h(x)\mathrm{d}x)\mathrm{d}t}\right]^\dagger \right. \tag{8}$$
$$\left.\mathcal{T}\left[\mathrm{e}^{-\mathrm{i}\int(H_0+\epsilon(t)\int\pi(x)h(x)\mathrm{d}x)\mathrm{d}t}\right]\right)\langle-|\frac{1}{2}(1+\boldsymbol{\rho}(0)\cdot\boldsymbol{\sigma})|+\rangle$$
$$= (\rho_1(0)+i\rho_2(0))J(t),$$

where H_0 is the Hamiltonian with $\epsilon = 0$, i.e., the free Hamiltonian for the scalar field and

$$J(t) = \tag{9}$$
$$\mathrm{Tr}_\phi\left(\mathcal{T}\left[\mathrm{e}^{-\mathrm{i}\int(H_0-\epsilon(t)\int\pi(x)h(x)\mathrm{d}x)\mathrm{d}t}\right]^\dagger\mathcal{T}\left[\mathrm{e}^{-\mathrm{i}\int(H_0+\epsilon(t)\int\pi(x)h(x)\mathrm{d}x)\mathrm{d}t}\right]R_0\right)$$

Breaking up the time ordered product in the standard way into a large number of small time steps, using the fact that $\exp[-\mathrm{i}\epsilon(t)\int h(x)\phi(x)\mathrm{d}x]$ is the displacement operator for the field momentum through a distance of $\epsilon(t)h(x)$, and commuting the free field Hamiltonian terms through, this can be written as

$$J(t) = \mathrm{Tr}_\phi\left(\mathrm{e}^{-\mathrm{i}\epsilon(0)\Phi(0)}\prod_{n=1}^{t/\mathrm{d}t}\left[\mathrm{e}^{-\mathrm{i}(\epsilon(t_n)-\epsilon(t_{n-1})\Phi(t_n))}\right]\right.$$
$$\left.\mathrm{e}^{i\epsilon(t)\Phi(t)}\mathrm{e}^{i\epsilon(t)\Phi(t)}\prod_{n=1}^{t/\mathrm{d}t}\left[\mathrm{e}^{i\epsilon(t_n-\epsilon(t_{n-1}))\Phi(t_n)}\right]\mathrm{e}^{i\epsilon(0)\Phi(0)}R_0\right), \tag{10}$$

where $t_n = n\mathrm{d}t$ and $\mathrm{d}t$ is a very small value, $\Phi(t) = \int h(x)\phi(t,x)\mathrm{d}x$ and $\phi_0(t,x)$ is the free field Heisenberg field operator. Using the Campbell-Baker-Hausdorff formula, realizing that the commutators of the Φs are c-numbers, and noticing that these c-numbers cancel between the two products, we finally get

$$J(t) = \mathrm{Tr}_\phi\left(\mathrm{e}^{2\mathrm{i}(\epsilon(t)\Phi(t)-\epsilon(0)\Phi(0)+\int_0^t\dot\epsilon(t')\Phi(t')\mathrm{d}t')}R_0\right) \tag{11}$$

from which we get

$$\ln(J(t)) = -2\mathrm{Tr}_\phi \left(R_0 \left(\epsilon(t)\Phi(t) - \epsilon(0)\Phi(0) + \int_0^t \dot{\epsilon}(t')\Phi(t')\mathrm{d}t' \right)^2 \right). \quad (12)$$

I will assume that $\epsilon(0) = 0$, and that $\dot{\epsilon}(t)$ is very small, and that it can be neglected. (The neglected terms are of the form

$$\int\int \dot{\epsilon}^2 \langle \Phi(t')\Phi(t'')\rangle \mathrm{d}t'\mathrm{d}t'' \approx \dot{\epsilon}^2 t\tau \langle \Phi(0)^2 \rangle$$

which for a massive scalar field has the coherence time scale $\tau \approx 1/m$. Thus, as we let $\dot{\epsilon}$ go to zero these terms go to zero.)

We finally have

$$\begin{aligned} \ln(J(t)) &= -2\epsilon(t)^2 < \Phi(t)^2 > \\ &= -2\epsilon(t)^2 \int |\hat{h}(k)|^2 \frac{1}{\sqrt{k^2+m^2}} \mathrm{d}k. \end{aligned} \quad (13)$$

Choosing $\hat{h}(k) = \mathrm{e}^{-\Gamma|k|/2}$, we finally get

$$\ln(J(t)) = -4 \int_0^\infty \epsilon(t)^2 \frac{\mathrm{e}^{-\Gamma|k|}\mathrm{d}k}{\sqrt{(k^2+m^2)}}. \quad (14)$$

This goes roughly as $\ln(\Gamma m)$ for small Γm, (which I will assume is true). For Γ sufficiently small, this makes J very small, and the density matrix reduces to essentially diagonal form ($\rho_z(t) \approx \rho_y(t) \approx 0$, $\rho_z(t) = \rho_z(0)$.)

However it is clear that if $\epsilon(t)$ is now lowered slowly to zero, the decoherence factor J goes back to unity, since it depends only on $\epsilon(t)$. The density matrix now has exactly its initial form again. The loss of coherence at the intermediate times was illusionary. By decoupling the system from the environment after the coherence had been lost, the coherence is restored. This is in contrast with the naive expectation in which the loss of coherence comes about because of the correlations between the system and the environment. Decoupling the system from the environment should not in itself destroy that correlation, and should not reestablish the coherence.

The above approach, while giving the correct results, is not very transparent in explaining what is happening. Let us therefore take a different approach. Let us solve the Heisenberg equations of motion for the field $\phi(t,x)$. The equations are (after eliminating π)

$$\partial_t^2\phi(t,x) - \partial_x^2\phi(t,x) + m^2\phi(t,x) = -\dot{\epsilon}(t)\sigma_3 h(x), \quad (15)$$

$$\pi(t,x) = \dot{\phi}(t,x) + \epsilon(t)h(x)\sigma_3. \quad (16)$$

If ϵ is slowly varying in time, we can solve this approximately by

$$\phi(t,x) = \phi_0(t,x) + \dot{\epsilon}(t) \int \frac{1}{2m} \mathrm{e}^{-m|x-x'|} h(x')\mathrm{d}x'\sigma_3 + \psi(t,x)\epsilon(0)\sigma_3, \quad (17)$$

$$\pi(t,x) = \dot{\phi}_0(t,x) + \epsilon(t)h(x)\sigma_3 + \dot{\psi}(t,x)\epsilon(0)\sigma_3, \quad (18)$$

where $\phi_0(t,x)$ and $\pi_0(t,x)$ are free field solution to the equations of motion in absence of the coupling, with the same initial conditions

$$\dot{\phi}_0(0,x) = \pi(0,x), \tag{19}$$
$$\phi_0(0,x) = \phi(0,x), \tag{20}$$

while ψ is also a solution of the free field equations but with initial conditions

$$\psi(0,x) = 0, \tag{21}$$
$$\dot{\psi}(0,x) = -h(x). \tag{22}$$

If we examine this for the two possible eigenstates of σ_3, we find the two solutions

$$\phi_\pm(t,x) \approx \phi_0(t,x) \pm \left(\dot{\epsilon}(t)\int \frac{1}{2m}\mathrm{e}^{-m|x-x'|}h(x')\mathrm{d}x' + \psi(t,x)\right), \tag{23}$$
$$\pi_\pm(t,x) \approx \dot{\phi}_0(t,x) + O(\dot{\epsilon}) \pm (\epsilon(t)h(x) + \epsilon(0)\dot{\psi}(t,x)). \tag{24}$$

These solutions neglect terms of higher derivatives in ϵ. The state of the field is the vacuum state of ϕ_0, π_0. $\phi_\pm$ and $\pi_\pm$ are equal to this initial field plus c-number fields. Thus in terms of the $\phi_\pm$ and $\pi_\pm$, the state is a coherent state with non-trivial displacement from the vacuum. Writing the fields in terms of their creation and annihilation operators,

$$\phi_\pm(t,x) = \int A_{k\pm}(t)\mathrm{e}^{\mathrm{i}kx} + A^\dagger_{k\pm}\mathrm{e}^{-\mathrm{i}kx}\frac{\mathrm{d}k}{\sqrt{2\pi\omega_k}}, \tag{25}$$
$$\pi_\pm(t,x) = \mathrm{i}\int A_{k\pm}(t)\mathrm{e}^{\mathrm{i}kx} - A^\dagger_{k\pm}\mathrm{e}^{-\mathrm{i}kx}\sqrt{\frac{k^2+m^2}{2\pi}}\mathrm{d}k, \tag{26}$$

we find that we can write $A_{k\pm}$ in terms of the initial operators A_{k0} as

$$A_{k\pm}(t) \approx A_{k0}\mathrm{e}^{-\mathrm{i}\omega_k t} \pm \frac{1}{2}\mathrm{i}(\epsilon(t) - \epsilon(0)\mathrm{e}^{-\mathrm{i}\omega_k t})(h(k)/\sqrt{\omega_k} + O(\dot{\epsilon}(t))), \tag{27}$$

where $\omega_k = \sqrt{k^2+m^2}$. Again I will neglect the terms of order $\dot{\epsilon}$ in comparison with the ϵ terms. Since the state is the vacuum state with respect to the initial operators A_{k0}, it will be a coherent state with respect to the operators $A_{k\pm}$, the annihilation operators for the field at time t. We thus have two possible coherent states for the field, depending on whether the spin is in the upper or lower eigenstate of σ_3. But these two coherent states will have a small overlap. If $A|\alpha\rangle = \alpha|\alpha\rangle$ then we have

$$|\alpha\rangle = \mathrm{e}^{\alpha A^\dagger - |\alpha|^2/2}|0\rangle. \tag{28}$$

Furthermore, if we have two coherent states $|\alpha\rangle$ and $|\alpha'\rangle$, then the overlap is given by

$$\langle\alpha|\alpha'\rangle = \langle 0|\mathrm{e}^{\alpha^* A - |\alpha|^2/2}\mathrm{e}^{\beta A^\dagger - |\beta|^2/2}|0\rangle = \mathrm{e}^{\alpha^*\beta - (|\alpha|^2+|\beta|^2)/2}. \tag{29}$$

In our case, taking the two states $|\pm_\phi\rangle$, these correspond to coherent states with

$$\alpha = -\alpha' = \frac{1}{2}\mathrm{i}(\epsilon(t) - \epsilon(0)\mathrm{e}^{-\mathrm{i}\omega_k t}) = \frac{1}{2}\mathrm{i}\epsilon(t)h(k)/\sqrt{\omega_k}. \tag{30}$$

Thus we have

$$< +_\phi, t|-_\phi, t >= \prod_k \mathrm{e}^{-\epsilon(t)^2|h(k)|^2/(k^2+m^2)} = \mathrm{e}^{-\epsilon(t)^2 \int \frac{|h(k)|^2}{\omega_k} dk} = J(t). \tag{31}$$

Let us assume that we began with the state of the spin as $\frac{1}{\sqrt{2}}(|+\rangle + |-\rangle)$. The state of the system at time t in the Schrödinger representation is

$$\frac{1}{\sqrt{2}}\left(|+\rangle| +_\phi (t)\rangle + |-\rangle|-_\phi\rangle\right)$$

and the reduced density matrix is

$$\rho = \frac{1}{2}(|+\rangle\langle+| + |-\rangle\langle-| + J^*(t)|+\rangle\langle-| + J(t)|-\rangle\langle+|). \tag{32}$$

The off diagonal terms of the density matrix are suppressed by the function $J(t)$. $J(t)$ however depends only on $\epsilon(t)$ and thus , as long as we keep $\dot{\epsilon}$ small, the loss of coherence represented by J can be reversed simply by decoupling the system from the environment slowly.

The apparent decoherence comes about precisely because the system in either the two eigenstates of σ_3 drives the field into two different coherent states. For large ϵ, these two states have small overlap. However, this distortion of the state of the field is tied to the system. π changes only locally, and the changes in the field caused by the system do not radiate away. As ϵ slowly changes, this bound state of the field also slowly changes in concert. However if one examines only the system, one sees a loss of coherence because the field states have only a small overlap with each other.

The behaviour is very different if the system or the interaction changes rapidly. In that case the decoherence can become real. As an example, consider the above case in which $\epsilon(t)$ suddenly is reduced to zero. In that case, the field is left as a free field, but a free field whose state (the coherent state) depends on the state of the system. In this case the field radiates away as real (not bound) excitations of the scalar field. The correlations with the system are carried away, and even if the coupling were again turned on, the loss of coherence would be permanent.

3 Oscillator

For the harmonic oscillator coupled to a heat bath, the Hamiltonian can be taken as

$$H = \frac{1}{2}\int[(\pi(x) - \epsilon(t)q(t)\tilde{h}(x))^2 + (\partial_x\phi(x))^2 + m^2\phi(t,x)^2]\mathrm{d}x + \frac{1}{2}(p^2 + \Omega^2 q^2). \tag{33}$$

Let us assume that m is much larger than Ω or the inverse timescale of change of ϵ. The solution for the field is given by

$$\phi(t,x) \approx \phi_0(t,x) + \psi(t,x)\epsilon(0)q(0) - \overline{\epsilon(t)q(t)}^{\cdot} \int \frac{\mathrm{e}^{-m|x-x'|}}{2m} h(x')\mathrm{d}x', \quad (34)$$

$$\pi(t,x) \approx \dot\phi_0(t,x) + \dot\psi(t,x)\epsilon(0)q(0) - \overline{\epsilon(t)q(t)}^{\cdot\cdot} \int \frac{\mathrm{e}^{-m|x-x'|}}{2m} h(x')\mathrm{d}x' + \epsilon(t)q(t)h(x), \quad (35)$$

where again ϕ_0 is the free field operator, ψ is a free field solution with $\psi(0) = 0$, $\dot\psi(0) = -h(x)$. Retaining terms only of the lowest order in ϵ,

$$\phi(t,x) \approx \phi_0(t,x), \quad (36)$$

$$\pi(t,x) \approx \dot\phi_0(t,x) + \epsilon(t)q(t)h(x). \quad (37)$$

The equation of motion for q is

$$\dot q(t) = p(t), \quad (38)$$

$$\dot p(t) = -\Omega^2 q + \epsilon(t)\dot\Phi(t), \quad (39)$$

where $\Phi(t) = \int h(x)\phi(t,x)\mathrm{d}x$. Substitution in the expression for ϕ, we get

$$\ddot q(t) + \Omega^2 q(t) \approx \epsilon(t)\dot\Phi_0(t) - \epsilon(t)\overline{\epsilon(t)q(t)}^{\cdot\cdot} \int\int h(x)h(x')\frac{\mathrm{e}^{-m|x-x'|}}{2m}\mathrm{d}x\mathrm{d}x'. \quad (40)$$

Neglecting the derivatives of ϵ (i.e., assuming that ϵ changes slowly even on the time scale of $1/\Omega$), this becomes

$$\left(1 + \epsilon(t)^2 \int\int h(x)h(x')\frac{\mathrm{e}^{-m|x-x'|}}{2m}\mathrm{d}x\mathrm{d}x'\right)\ddot q + \Omega^2 q = \partial_t(\epsilon(t)\Phi(t)). \quad (41)$$

The interaction with the field thus renormalizes the mass of the oscillator to

$$M = 1 + \epsilon(t)^2 \int\int h(x)h(x')\frac{\mathrm{e}^{-m|x-x'|}}{2m}\mathrm{d}x\mathrm{d}x'.$$

The solution for q is thus

$$q(t) \approx q(0)\cos\left(\int_0^t \tilde\Omega(t)\mathrm{d}t\right) + \frac{1}{\tilde\Omega}\sin\left(\int_0^t \tilde\Omega(t)\mathrm{d}t\right)p(0) + \frac{1}{\tilde\Omega}\int_0^t \sin\left(\int_{t'}^t \tilde\Omega(t)\mathrm{d}t\right)\partial_t(\epsilon(t')\overline{\epsilon(t)\Phi_0(t')}^{\cdot}\mathrm{d}t', \quad (42)$$

where $\tilde\Omega(t) \approx \Omega/\sqrt{M(t)}$.

The important point is that the forcing term dependent on Φ_0 is a rapidly oscillating term of frequency at least m. Thus if we look for example at $\langle q^2\rangle$, the deviation from the free evolution of the oscillator (with the renormalized mass) is of the order of

$$\int \sin(\tilde{\Omega}t - t') \sin(\omega(t - t''))\langle \dot{\Phi}_0(t')\dot{\Phi}_0(t'')\rangle dt' dt''.$$

But $\langle \dot{\Phi}_0(t')\dot{\Phi}_0(t'')\rangle$ is a rapidly oscillating function of frequency at least m, while the rest of the integrand is a slowly varying function with frequency much less than m. Thus this integral will be very small (at least $\tilde{\Omega}/m$ but typically much smaller than this depending on the time dependence of ϵ). Thus the deviation of $q(t)$ from the free motion will in general be very very small, and I will neglect it.

Let us now look at the field. The field is put into a coherent state which depends on the value of q, because $\pi(t,x) \approx \dot{\phi}_0(t,x) + \epsilon(t)q(t)h(x)$. Thus,

$$A_k(t) \approx a_{0k}e^{-i\omega_k t} + i\frac{1}{2}\hat{h}(k)\epsilon(t)q(t)/\omega_k. \tag{43}$$

The overlap integral for these coherent states with various values of q is

$$\prod_k \langle i\frac{1}{2}\hat{h}(k)\epsilon(t)q/\omega_k | i\frac{1}{2}\hat{h}(k)\epsilon(t)q'/\omega_k\rangle = e^{-\frac{1}{8}\int |\hat{h}(k)|^2 dk (q-q')^2}. \tag{44}$$

The density matrix for the Harmonic oscillator is thus

$$\rho(q,q') = \rho_0(t,q,q')e^{-\frac{1}{8}\int |\hat{h}(k)|^2 dk(q-q')^2}, \tag{45}$$

where ρ_0 is the density matrix for a free harmonic oscillator (with the renormalized mass).

We see a strong loss of coherence of the off diagonal terms of the density matrix. However this loss of coherence is false. If we take the initial state for example with two packets widely separated in space, these two packets will loose their coherence. However, as time proceeds, the natural evolution of the Harmonic oscillator will bring those two packets together ($q - q'$ small across the wave packet). For the free evolution they would then interfere. They still do. The loss of coherence which was apparent when the two packets were widely separated disappears, and the two packets interfere just as if there were no coupling to the environment. The effect of the particular environment used is thus to renormalise the mass, and to make the density matrix appear to loose coherence.

4 Spin Boson Problem

Let us now complicate the spin problem in the first section by introducing into the system a free Hamiltonian for the spin as well as the coupling to the environment. Following the example of the spin boson problem, let me introduce a free Hamiltonian for the spin of the form $\frac{1}{2}\Omega\sigma_1$, whose effect is to rotate the σ_3 states (or to rotate the vector $\boldsymbol{\rho}$ in the $2-3$ plane) with frequency Ω.

The Hamiltonian now is

$$H = \frac{1}{2}\left(\int[(\pi(t,x) - \epsilon(t)h(x)\sigma_3)^2 + (\partial_x\phi(x))^2 + m^2\phi(t,x)^2]\mathrm{d}x + \Omega\sigma_1\right), \tag{46}$$

where again $\epsilon(t)$ is a slowly varying function of time. We will solve this in the manner of the second part of section 2.

If we let Ω be zero, then the eigenstates of σ_z are eigenstates of the Hamiltonian. The field Hamiltonian (for constant ϵ) is given by

$$H_\pm = \frac{1}{2}\int[(\pi - (\pm\epsilon(t)h(x)))^2 + (\partial_x\phi)^2]\mathrm{d}x. \tag{47}$$

Defining $\tilde{\pi} = \pi - (\pm h(x))$, $\tilde{\pi}$ has the same commutation relations with π and ϕ as does π. Thus in terms of $\tilde{\pi}$ we just have the Hamiltonian for the free scalar field. The instantaneous minimum energy state is therefore the ground state energy for the free scalar field for both $H_\pm$. Thus the two states are degenerate in energy. In terms of the operators π and ϕ, these ground states are coherent states with respect to the vacuum state of the original uncoupled ($\epsilon = 0$) free field, with the displacement of each mode given by

$$a_k|\pm\rangle = \pm\mathrm{i}\epsilon(t)\frac{h(k)}{\sqrt{\omega_k}}|\pm\rangle, \tag{48}$$

or

$$|\pm\rangle = \prod_k |\pm\alpha_k\rangle|\pm\rangle_{\sigma_3}, \tag{49}$$

where the $|\alpha_k\rangle$ are coherent states for the kth modes with coherence parameter $\alpha_k = \mathrm{i}\epsilon(t)\frac{h(k)}{\sqrt{\omega_k}}$, and the states $|\pm\rangle_{\sigma_3}$ are the two eigenstates of σ_3. (In the following I will eliminate the $\prod_k$ symbol.) The energy to the next excited state in each case is just m, the mass of the free field.

We now introduce the $\Omega\sigma_x$ as a perturbation parameter. The two lowest states (and in fact the excited states) are two-fold degenerate. Using degenerate perturbation theory to find the new lowest energy eigenstates, we must calculate the overlap integral of the perturbation between the original degenerate states and must then diagonalise the resultant matrix to lowest order in Ω. The perturbation is $\frac{1}{2}\Omega\sigma_1$. All terms between the same states are zero, because of the $\langle\pm|_{\sigma_3}\sigma_1|\pm\rangle_{\sigma_3} = 0$. Thus the only terms that survive for determining the lowest order correction to the lowest energy eigenvalues are

$$\frac{1}{2}\langle +|\Omega\sigma_1|-\rangle = \frac{1}{2}\langle -|\Omega\sigma_1|+\rangle^* \tag{50}$$

$$= \frac{1}{2}\Omega\prod_k\langle\alpha_k| - \alpha_k\rangle = \frac{1}{2}\Omega\prod_k \mathrm{e}^{-2|\alpha_k|^2} \tag{51}$$

$$= \frac{1}{2}\Omega \mathrm{e}^{-2\int \epsilon(t)^2 |h(k)|^2/\omega_k \mathrm{d}k} = \frac{1}{2}\Omega J(t). \tag{52}$$

The eigenstates of energy thus have energy of $E(t)_\pm = E_0 \pm \frac{1}{2}\Omega J(t)$, and the eigenstates are $\sqrt{\frac{1}{2}}(|+\rangle \pm |-\rangle)$. If ϵ varies slowly enough, the instantaneous energy eigenstates will be the actual adiabatic eigenstates at all times, and the time evolution of the system will just be in terms of these instantaneous energy eigenstates. Thus the system will evolve as

$$|\psi(t)\rangle = \sqrt{\tfrac{1}{2}}\mathrm{e}^{-\mathrm{i}E_0 t}\Big((c_+ + c_-)\mathrm{e}^{-\mathrm{i}\int \frac{1}{2}\Omega_t J(t)\mathrm{d}t}(|+\rangle + |-\rangle) \\ + (c_- - c_+)\mathrm{e}^{+\mathrm{i}\int \frac{1}{2}\Omega_t J(t)\mathrm{d}t}(|+\rangle - |-\rangle)\Big), \tag{53}$$

where the c_+ and c_- are the initial amplitudes for the $|+\rangle_{\sigma_3}$ and $|-\rangle_{\sigma_3}$ states. The reduced density matrix for the spin system in the σ_3 basis can now be written as

$$\boldsymbol{\rho}(t) = (J(t)\rho_{01}(t), J(t)\rho_{02}(t), \rho_{03}(t)), \tag{54}$$

where $\boldsymbol{\rho}_0(t)$ is the density matrix that one would obtain for a free spin half particle moving under the Hamiltonian $J(t)\Omega\sigma_1$,

$$\begin{aligned} \rho_{01}(t) &= \rho_1(0), \\ \rho_{02}(t) &= \rho_2(0)\cos\left(\Omega\int J(t')\mathrm{d}t'\right) + \rho_3(0)\sin\left(\Omega\int J(t')\mathrm{d}t'\right), \\ \rho_{03}(t) &= \rho_3(0)\cos\left(\Omega\int J(t')\mathrm{d}t'\right) - \rho_2(0)\sin\left(\Omega\int J(t')\mathrm{d}t\right). \end{aligned} \tag{55}$$

Thus, if $J(t)$ is very small (i.e., ϵ large), we have a renormalized frequency for the spin system, and the the off diagonal terms (in the σ_3 representation) of the density matrix are strongly suppressed by a factor of $J(t)$. Thus if we begin in an eigenstate of σ_3 the density matrix will begin with the vector $\boldsymbol{\rho}$ as a unit vector pointing in the 3 direction. As time goes on the 3 component gradually decreases to zero, but the 2 component increases only to the small value of $J(t)$. The system looks almost like a completely incoherent state, with almost the maximal entropy that the spin system could have. However, as we wait longer, the 3 component of the density vector reappears and grows back to its full unit value in the opposite direction, and the entropy drop to zero again. This cycle repeats itself endlessly with the entropy oscillating between its minimum and maximum value forever.

The decoherence of the density matrix (the small off diagonal terms) obviously represent a false loss of coherence. It represents a strong correlation between the system and the environment. However the environment is bound to the system, and essentially forms a part of the system itself, at least as long as the system moves slowly. However the reduced density matrix makes no distinction between whether or not the correlations between the system and the environment are in some sense bound to the system, or are correlations between the system and a freely propagating modes of the medium in which case the correlations can be extremely difficult to recover, and certainly cannot be recovered purely by manipulations of the system alone.

5 Instantaneous Change

In the above I have assumed throughout that the system moves slowly with respect to the excitations of the heat bath. Let us now look at what happens in the spin system if we rapidly change the spin of the system. In particular I will assume that the system is as in section 1, a spin coupled only to the massive heat bath via the component σ_3 of the spin. Then at a time t_0, I instantly rotate the spin through some angle θ about the 1 axis. In this case we will find that the environment cannot adjust rapidly enough, and at least a part of the loss of coherence becomes real, becomes unrecoverable purely through manipulations of the spin alone.

The Hamiltonian is

$$H = \frac{1}{2}\int [(\pi(t,x) - \epsilon(t)h(x)\sigma_3)^2 + (\partial_x\phi(t,x))^2 + m^2\phi(t,x)]\mathrm{d}x \\ +\theta/2\delta(t-t_0)\sigma_1. \tag{56}$$

Until the time t_0 σ_3 is a constant of the motion, and similarly afterward. Before the time t_0, the energy eigenstates state of the system are as in the last section given by

$$|\pm,t\rangle = \{|+\rangle_{\sigma_3}|\alpha_k(t)\rangle \text{ or } \{|-\rangle_{\sigma_3}|-\alpha_k(t)\rangle\}. \tag{57}$$

An arbitrary state for the spin–environment system is given by

$$|\psi\rangle = c_+|+\rangle + c_-|-\rangle. \tag{58}$$

Now, at time t_0, the rotation carries this to

$$\begin{aligned} |\phi(t_0)\rangle &= c_+(\cos(\theta/2)|+\rangle_{\sigma_3} + \mathrm{i}\sin(\theta/2)|-\rangle_{\sigma_3})|\alpha_k(t)\rangle \\ &\quad + c_-(\cos(\theta/2)|-\rangle_{\sigma_3} + \mathrm{i}\sin(\theta/2)|+\rangle_{\sigma_3})|-\alpha_k(t)\rangle \\ &= \cos(\theta/2)\,(c_+|+\rangle + c_-|-\rangle) \\ &\quad + \mathrm{i}\sin(\theta/2)(c_+|-\rangle_{\sigma_3}|\alpha_k(t)\rangle - c_-|+\rangle_{\sigma_3}|-\alpha_k(t)\rangle). \end{aligned} \tag{59}$$

The first term is still a simple sum of eigenvectors of the Hamiltonian after the interaction. The second term, however, is not. We thus need to follow the evolution of the two states $|-\rangle_{\sigma_3}|\alpha_k(t_0)\rangle$ and $|+\rangle_{\sigma_3}|-\alpha_k(t_0)\rangle$. Since σ_3 is a constant of the motion after the interaction again, the evolution takes place completely in the field sector. Let us look at the first state first. (The evolution of the second can be derived easily from that for the first because of the symmetry of the problem.)

I will again work in the Heisenberg representation. The field obeys

$$\dot{\phi}_-(t,x) = \pi_-(t,x) + \epsilon(t)h(x), \tag{60}$$
$$\dot{\pi}_-(t,x) = \partial_x^2\phi_-(t,x) - m^2\phi_-(t,x). \tag{61}$$

At the time t_0 the field is in the coherent state $|\alpha_k\rangle$. This can be represented by taking the field operator to be of the form

$$\phi_-(t_0,x) = \phi_0(t_0,x), \tag{62}$$
$$\pi_-(t_0,x) = \dot{\phi}_0(t_0,x) + \epsilon(t_0)h(x), \tag{63}$$

where the state $|\alpha_k\rangle$ is the vacuum state for the free field ϕ_0. We can now solve the equations of motion for ϕ_- and obtain (again assuming that $\epsilon(t)$ is slowly varying)

$$\phi_-(t,x) = \phi_0(t,x) + 2\psi(t,x)\epsilon(t_0), \tag{64}$$
$$\pi_-(t,x) = \dot{\phi}_0(t,x) + 2\dot{\psi}(t,x)\epsilon(t_0) - \epsilon(t)h(x), \tag{65}$$

where $\psi(t_0,x) = 0$ and $\dot{\psi}(t_0,x) = h(x)$. Thus again, the field is in a coherent state set by both $2\epsilon(t_0)\psi$ and $\epsilon(t)h(x)$. The field ψ propagates away from the interaction region determined by $h(x)$, and I will assume that I am interested in times t a long time after the time t_0. At these times I will assume that $\int h(x)\psi(t,x)\mathrm{d}x = 0$. (This overlap dies out as $1/\sqrt{mt}$. The calculations can be carried out for times nearer t_0 as well— the expressions are just messier and not particularly informative.)

Let me define the new coherent state as $|-\alpha_k(t) + \beta_k(t)\rangle$, where α_k is as before and

$$\beta_k(t) = 2\epsilon(t_0)\omega_k\tilde{\psi}(t,k) = 2\mathrm{i}\epsilon(t_0)\mathrm{e}^{\mathrm{i}\omega_k t}\tilde{h}(k)/\omega_k. \tag{66}$$

(The assumption regarding the overlap of $h(x)$ and $\psi(t)$ corresponds to the assumption that $\int \alpha_k^*(t)\beta_k(t)\mathrm{d}k = 0$). Thus the state $|-\rangle_{\sigma 3}|\alpha_k\rangle$ evolves to the state $|-\rangle_{\sigma 3}|-\alpha_k + \beta_k(t)\rangle$. Similarly, the state $|+\rangle_{\sigma 3}|-\alpha_k\rangle$ evolves to $|+\rangle_{\sigma 3}|\alpha_k - \beta_k(t)\rangle$.)

We now calculate the overlaps of the various states of interest.

$$\langle\alpha_k|\alpha_k \pm \beta_k\rangle = \langle-\alpha_k| - \alpha_k \pm \beta_k\rangle = \mathrm{e}^{-\int|\beta_k|^2\mathrm{d}k} = J(t_0), \tag{67}$$
$$\langle-\alpha_k|\alpha_k \pm \beta_k\rangle = \langle\alpha_k| - \alpha_k \pm \beta_k\rangle = J(t)J(t_0), \tag{68}$$
$$\langle-\alpha_k + \beta_k|\alpha_k - \beta_k\rangle = \langle-\alpha_k - \beta_k|\alpha_k + \beta_k\rangle = J(t)J(t_0)^4. \tag{69}$$

The density matrix becomes

$$\rho_3 = \cos(\theta)\rho_{03} + \sin(\theta)J(t_0)\rho_{02}, \tag{70}$$
$$\rho_1 = J(t)\left(\cos(\theta) + J^4(t_0)\sin(\theta)\right)\rho_{01}, \tag{71}$$
$$\rho_2(t) = J(t)\left(-\sin(\theta)\rho_{03} + (\cos(\theta/2) - J^4(t_0)\sin(\theta))\rho_{02}\right), \tag{72}$$

where

$$\rho_{03} = \frac{1}{2}(|c_+|^2 - |c_-|^2), \tag{73}$$

$$\rho_{01} = \mathrm{Re}(c_+ c_-^*), \tag{74}$$

$$\rho_{02} = \mathrm{Im}(c_+ c_-^*). \tag{75}$$

If we now let $\epsilon(t)$ go slowly to zero again (to find the 'real' loss of coherence), we find that unless $\rho_{01} = \rho_{02} = 0$ the system has really lost coherence during the sudden transition. The maximum real loss of coherence occurs if the rotation is a spin flip ($\theta = \pi$) and ρ_{03} was zero. In that case the density vector dropped to $J(t_0)^4$ of its original value. If the density matrix was in an eigenstate of σ_3 on the other hand, the density matrix remained a coherent density matrix, but the environment was still excited by the spin.

We can use the models of a fast or a slow spin flip interaction to discuss the problem of the tunneling time. As Leggett et al. argue [3], the spin system is a good model for the consideration of the behaviour of a particle in two wells, with a tunneling barrier between the two wells. One view of the transition from one well to the other is that the particle sits in one well for a long time. Then at some random time it suddenly jumps through the barrier to the other side. An alternative view would be to see the particle as if it were a fluid, with a narrow pipe connecting it to the other well- the fluid slowly sloshing between the two wells. The former is supported by the fact that if one periodically observes which of the two wells the particle is in, one sees it staying in one well for a long time, and then between two observations, suddenly finding it in the other well. This would, if one regarded it as a classical particle imply that the whole tunneling must have occurred between the two observations. It is as if the system were in an eigenstate and at some random time an interaction flipped the particle from one well to the other. However, this is not a good picture. The environment is continually observing the system. It is really moved rapidly from one to the other, the environment would see the rapid change, and would radiate. Instead, left on its own, the environment in this problem (with a mass much greater than the frequency of transition of the system) simply adjust continually to the changes in the system. The tunneling thus seems to take place continually and slowly.

6 Discussion

The high frequency modes of the environment lead to a loss of coherence (decay of the off-diagonal terms in the density matrix) of the system, but as long as the changes in the system are slow enough this decoherence is false– it does not prevent the quantum interference of the system. The reason is that the changes in the environment caused by these modes are essentially tied to the system, they are adiabatic changes to the environment which can easily be adiabatically reversed. Loosely one can say that coherence is lost by the transfer of information (coherence) from the system to the environment. However in order for this information to be truly lost, it must be carried away by the environment, separated from the system by some mechanism or another so that it cannot come back into the system. In the environment above, this occurs when the information travels off to infinity. Thus the loss of coherence as represented by the reduced density matrix is in some sense the maximum loss of coherence of the system. Rapid changes to the system, or rapid decoupling of the system from the environment, will make this a true decoherence. However, gradual changes in the system or in the coupling to the external world can cause the environment to adiabatically track the system and restore the coherence apparently lost.

This is of special importance to understanding the effects of the environmental cutoff in many environments [3]. For "ohmic" or "superohmic" environments (where h does not fall off for large arguments), one has to introduce a cutoff into the calculation for the reduced density matrix. This cutoff has always been a bit mysterious, especially as the loss of coherence depends sensitively on the value of this cutoff. If one imagines the environment to include say the electromagnetic field, what is the right value for this cutoff? Choosing the Plank scale seems silly, but what is the proper value? The arguments of this paper suggest that in fact the cutoff is unnecessary except in renormalising the dynamics of the system. The behaviour of the environment at frequencies much higher than the inverse time scale of the system leads to a false loss of coherence, a loss of coherence which does not affect the actual coherence (ability to interfere with itself) of the system. Thus the true coherence is independent of cutoff.

As far as the system itself is concerned, one should regard it as "dressed" with a polarization of the high frequency components of the environment. One should regard not the system itself as important for the quantum coherence, but a combination of variables of the system plus the environment.What is difficult is the question as to which degrees of freedom of the environment are simply dressing and which degrees of freedom can lead to loss of coherence. This question depends crucially on the motion and the interactions of the system itself. They are history dependent, not simply state dependent. This makes it very difficult to simply find some transformation which will express the system plus environment in terms of variables which are genuinely independent, in the sense that if the new variable loose coherence, then that loss is real.

These observations emphasise the importance of not making too rapid conclusions from the decoherence of the system. This is especially true in cosmology, where high frequency modes of the cosmological system are used to decohere low frequency quantum modes of the universe. Those high frequency modes are likely to behave adiabatically with respect to the low frequency behaviour of the universe. Thus, although they will lead to a reduced density matrix for the low frequency modes which is apparently incoherent, that incoherence is likely to be a false loss of coherence.

Acknowledgements

I would like to thank the Canadian Institute for Advanced Research for their support of this research. This research was carried out under an NSERC grant 580441.

References

1. Caldeira A. O., Leggett A. J. (1983): Physica **121A**, 587; (1985) Phys Rev **A31** , 1057. See also the paper by W. Unruh W., Zurek W. (1989): Phys Rev **D40**, 1071 where a field model for coherence instead of the oscillator model for calculating the density matrix of an oscillator coupled to a heat bath.
2. Many of the points made here have also been made by A. Leggett. See for example Leggett A. J. (1990). In Baeriswyl D., Bishop A. R., Carmelo J. (Eds.) *Applications of Statistical and Field Theory Methods to Condensed Matter*, Proc. 1989 Nato Summer School, Evora, Portugal. Plenum Press and (1998) Macroscopic Realism: What is it, and What do we know about it from Experiment. In Healey R. A., Hellman G. (Eds.), *Quantum Measurement: Beyond Paradox*, U. Minnesota Press, Minneapolis.
3. See for example the detailed analysis of the density matrix of a spin 1/2 system in an oscillator heat bath, where the so called superohmic coupling to the heat bath leads to a rapid loss of coherence due to frequencies in the bath much higher than the frequency of the system under study. Leggett A. J. et al. (1987): Rev. Mod. Phys **59**, 1.
4. This topic is a long standing one. For a review see Landauer R. and Martin T. (1994): Reviews of Modern Physics **66**, 217.

Lecture Notes in Physics

For information about Vols. 1–523
please contact your bookseller or Springer-Verlag

Vol. 524: J. Wess, E. A. Ivanov (Eds.), Supersymmetries and Quantum Symmetries. Proceedings, 1997. XX, 442 pages. 1999.

Vol. 525: A. Ceresole, C. Kounnas, D. Lüst, S. Theisen (Eds.), Quantum Aspects of Gauge Theories, Supersymmetry and Unification. Proceedings, 1998. X, 511 pages. 1999.

Vol. 526: H.-P. Breuer, F. Petruccione (Eds.), Open Systems and Measurement in Relativistic Quantum Theory. Proceedings, 1998. VIII, 240 pages. 1999.

Vol. 527: D. Reguera, J. M. G. Vilar, J. M. Rubí (Eds.), Statistical Mechanics of Biocomplexity. Proceedings, 1998. XI, 318 pages. 1999.

Vol. 528: I. Peschel, X. Wang, M. Kaulke, K. Hallberg (Eds.), Density-Matrix Renormalization. Proceedings, 1998. XVI, 355 pages. 1999.

Vol. 529: S. Biringen, H. Örs, A. Tezel, J.H. Ferziger (Eds.), Industrial and Environmental Applications of Direct and Large-Eddy Simulation. Proceedings, 1998. XVI, 301 pages. 1999.

Vol. 530: H.-J. Röser, K. Meisenheimer (Eds.), The Radio Galaxy Messier 87. Proceedings, 1997. XIII, 342 pages. 1999.

Vol. 531: H. Benisty, J.-M. Gérard, R. Houdré, J. Rarity, C. Weisbuch (Eds.), Confined Photon Systems. Proceedings, 1998. X, 496 pages. 1999.

Vol. 532: S. C. Müller, J. Parisi, W. Zimmermann (Eds.), Transport and Structure. Their Competitive Roles in Biophysics and Chemistry. XII, 400 pages. 1999.

Vol. 533: K. Hutter, Y. Wang, H. Beer (Eds.), Advances in Cold-Region Thermal Engineering and Sciences. Proceedings, 1999. XIV, 608 pages. 1999.

Vol. 534: F. Moreno, F. González (Eds.), Light Scattering from Microstructures. Proceedings, 1998. XII, 300 pages. 2000

Vol. 535: H. Dreyssé (Ed.), Electronic Structure and Physical Properties of Solids: The Uses of the LMTO Method. Proceedings, 1998. XIV, 458 pages. 2000.

Vol. 536: T. Passot, P.-L. Sulem (Eds.), Nonlinear MHD Waves and Turbulence. Proceedings, 1998. X, 385 pages. 1999.

Vol. 537: S. Cotsakis, G. W. Gibbons (Eds.), Mathematical and Quantum Aspects of Relativity and Cosmology. Proceedings, 1998. XII, 251 pages. 1999.

Vol. 538: Ph. Blanchard, D. Giulini, E. Joos, C. Kiefer, I.-O. Stamatescu (Eds.), Decoherence: Theoretical, Experimental, and Conceptual Problems. Proceedings, 1998. XII, 345 pages. 2000.

Vol. 539: A. Borowiec, W. Cegła, B. Jancewicz, W. Karwowski (Eds.), Theoretical Physics. Fin de Siècle. Proceedings, 1998. XX, 319 pages. 2000.

Vol. 540: B. G. Schmidt (Ed.), Einstein's Field Equations and Their Physical Implications. Selected Essays. 1999. XIII, 429 pages. 2000

Vol. 541: J. Kowalski-Glikman (Ed.), Towards Quantum Gravity. Proceedings, 1999. XII, 376 pages. 2000.

Vol. 542: P. L. Christiansen, M. P. Sørensen, A. C. Scott (Eds.), Nonlinear Science at the Dawn of the 21st Century. Proceedings, 1998. XXVI, 458 pages. 2000.

Vol. 543: H. Gausterer, H. Grosse, L. Pittner (Eds.), Geometry and Quantum Physics. Proceedings, 1999. VIII, 408 pages. 2000.

Vol. 544: T. Brandes (Ed.), Low-Dimensional Systems. Interactions and Transport Properties. Proceedings, 1999. VIII, 219 pages. 2000

Vol. 545: J. Klamut, B. W. Veal, B. M. Dabrowski, P. W. Klamut, M. Kazimierski (Eds.), New Developments in High-Temperature Superconductivity. Proceedings, 1998. VIII, 275 pages. 2000.

Vol. 546: G. Grindhammer, B. A. Kniehl, G. Kramer (Eds.), New Trends in HERA Physics 1999. Proceedings, 1999. XIV, 460 pages. 2000.

Vol. 547: D. Reguera, G. Platero, L.L. Bonilla, J.M. Rubí(Eds.), Statistical and Dynamical Aspects of Mesoscopic Systems. Proceedings, 1999. XII, 357 pages. 2000.

Vol. 548: D. Lemke, M. Stickel, K. Wilke (Eds.), ISO Surveys of a Dusty Universe. Proceedings, 1999. XIV, 432 pages. 2000.

Vol. 549: C. Egbers, G. Pfister (Eds.), Physics of Rotating Fluids. Proceedings, 1999. XVIII, 437 pages. 2000.

Vol. 550: M. Planat (Ed.), Noise, Oscillators and Algebraic Randomness. Proceedings, 1999. VIII, 417 pages. 2000.

Vol. 551: B. Brogliato (Ed.), Impacts in Mechanical Systems. Analysis and Modelling. Lectures, 1999. IX, 273 pages. 2000.

Vol. 552: Z. Chen, R. E. Ewing, Z.-C. Shi (Eds.), Numerical Treatment of Multiphase Flows in Porous Media. Proceedings, 1999. XXI, 445 pages. 2000.

Vol. 553: J.-P. Rozelot, L. Klein, J.-C. Vial Eds.), Transport of Energy Conversion in the Heliosphere. Proceedings, 1998. IX, 214 pages. 2000.

Vol. 554: K. R. Mecke, D. Stoyan (Eds.), Statistical Physics and Spatial Statistics. The Art of Analyzing and Modeling Spatial Structures and Pattern Formation. Proceedings, 1999. XII, 415 pages. 2000.

Vol. 555: A. Maurel, P. Petitjeans (Eds.), Vortex Structure and Dynamics. Proceedings, 1999. XII, 319 pages. 2000.

Vol. 556: D. Page, J. G. Hirsch (Eds.), GTO Lectures on Astrophysics. Proceedings, 1999. X, 330 pages. 2000.

Vol. 557: J. A. Freund, T. Pöschel (Eds.), Stochastic Processes in Physics, Chemistry, and Biology. X, 330 pages. 2000.

Vol. 558: P. Breitenlohner, D. Maison (Eds.), Quantum Field Theory. Proceedings, 1998. VIII, 323 pages. 2000

Vol. 559: H.-P. Breuer, F. Petruccione (Eds.), Relativistic Quantum Measurement and Decoherence. Proceedings, 1999. X, 140 pages. 2000.

Monographs

For information about Vols. 1–21
please contact your bookseller or Springer-Verlag

Vol. m 22: M.-O. Hongler, Chaotic and Stochastic Behaviour in Automatic Production Lines. V, 85 pages. 1994.

Vol. m 23: V. S. Viswanath, G. Müller, The Recursion Method. X, 259 pages. 1994.

Vol. m 24: A. Ern, V. Giovangigli, Multicomponent Transport Algorithms. XIV, 427 pages. 1994.

Vol. m 25: A. V. Bogdanov, G. V. Dubrovskiy, M. P. Krutikov, D. V. Kulginov, V. M. Strelchenya, Interaction of Gases with Surfaces. XIV, 132 pages. 1995.

Vol. m 26: M. Dineykhan, G. V. Efimov, G. Ganbold, S. N. Nedelko, Oscillator Representation in Quantum Physics. IX, 279 pages. 1995.

Vol. m 27: J. T. Ottesen, Infinite Dimensional Groups and Algebras in Quantum Physics. IX, 218 pages. 1995.

Vol. m 28: O. Piguet, S. P. Sorella, Algebraic Renormalization. IX, 134 pages. 1995.

Vol. m 29: C. Bendjaballah, Introduction to Photon Communication. VII, 193 pages. 1995.

Vol. m 30: A. J. Greer, W. J. Kossler, Low Magnetic Fields in Anisotropic Superconductors. VII, 161 pages. 1995.

Vol. m 31 (Corr. Second Printing): P. Busch, M. Grabowski, P.J. Lahti, Operational Quantum Physics. XII, 230 pages. 1997.

Vol. m 32: L. de Broglie, Diverses questions de mécanique et de thermodynamique classiques et relativistes. XII, 198 pages. 1995.

Vol. m 33: R. Alkofer, H. Reinhardt, Chiral Quark Dynamics. VIII, 115 pages. 1995.

Vol. m 34: R. Jost, Das Märchen vom Elfenbeinernen Turm. VIII, 286 pages. 1995.

Vol. m 35: E. Elizalde, Ten Physical Applications of Spectral Zeta Functions. XIV, 224 pages. 1995.

Vol. m 36: G. Dunne, Self-Dual Chern-Simons Theories. X, 217 pages. 1995.

Vol. m 37: S. Childress, A.D. Gilbert, Stretch, Twist, Fold: The Fast Dynamo. XI, 406 pages. 1995.

Vol. m 38: J. González, M. A. Martín-Delgado, G. Sierra, A. H. Vozmediano, Quantum Electron Liquids and High-Tc Superconductivity. X, 299 pages. 1995.

Vol. m 39: L. Pittner, Algebraic Foundations of Non-Com-mutative Differential Geometry and Quantum Groups. XII, 469 pages. 1996.

Vol. m 40: H.-J. Borchers, Translation Group and Particle Representations in Quantum Field Theory. VII, 131 pages. 1996.

Vol. m 41: B. K. Chakrabarti, A. Dutta, P. Sen, Quantum Ising Phases and Transitions in Transverse Ising Models. X, 204 pages. 1996.

Vol. m 42: P. Bouwknegt, J. McCarthy, K. Pilch, The W3 Algebra. Modules, Semi-infinite Cohomology and BV Algebras. XI, 204 pages. 1996.

Vol. m 43: M. Schottenloher, A Mathematical Introduction to Conformal Field Theory. VIII, 142 pages. 1997.

Vol. m 44: A. Bach, Indistinguishable Classical Particles. VIII, 157 pages. 1997.

Vol. m 45: M. Ferrari, V. T. Granik, A. Imam, J. C. Nadeau (Eds.), Advances in Doublet Mechanics. XVI, 214 pages. 1997.

Vol. m 46: M. Camenzind, Les noyaux actifs de galaxies. XVIII, 218 pages. 1997.

Vol. m 47: L. M. Zubov, Nonlinear Theory of Dislocations and Disclinations in Elastic Body. VI, 205 pages. 1997.

Vol. m 48: P. Kopietz, Bosonization of Interacting Fermions in Arbitrary Dimensions. XII, 259 pages. 1997.

Vol. m 49: M. Zak, J. B. Zbilut, R. E. Meyers, From Instability to Intelligence. Complexity and Predictability in Nonlinear Dynamics. XIV, 552 pages. 1997.

Vol. m 50: J. Ambjørn, M. Carfora, A. Marzuoli, The Geometry of Dynamical Triangulations. VI, 197 pages. 1997.

Vol. m 51: G. Landi, An Introduction to Noncommutative Spaces and Their Geometries. XI, 200 pages. 1997.

Vol. m 52: M. Hénon, Generating Families in the Restricted Three-Body Problem. XI, 278 pages. 1997.

Vol. m 53: M. Gad-el-Hak, A. Pollard, J.-P. Bonnet (Eds.), Flow Control. Fundamentals and Practices. XII, 527 pages. 1998.

Vol. m 54: Y. Suzuki, K. Varga, Stochastic Variational Approach to Quantum-Mechanical Few-Body Problems. XIV, 324 pages. 1998.

Vol. m 55: F. Busse, S. C. Müller, Evolution of Spontaneous Structures in Dissipative Continuous Systems. X, 559 pages. 1998.

Vol. m 56: R. Haussmann, Self-consistent Quantum Field Theory and Bosonization for Strongly Correlated Electron Systems. VIII, 173 pages. 1999.

Vol. m 57: G. Cicogna, G. Gaeta, Symmetry and Perturbation Theory in Nonlinear Dynamics. XI, 208 pages. 1999.

Vol. m 58: J. Daillant, A. Gibaud (Eds.), X-Ray and Neutron Reflectivity: Principles and Applications. XVIII, 331 pages. 1999.

Vol. m 59: M. Kriele, Spacetime. Foundations of General Relativity and Differential Geometry. XV, 432 pages. 1999.

Vol. m 60: J. T. Londergan, J. P. Carini, D. P. Murdock, Binding and Scattering in Two-Dimensional Systems. Applications to Quantum Wires, Waveguides and Photonic Crystals. X, 222 pages. 1999.

Vol. m 61: V. Perlick, Ray Optics, Fermat's Principle, and Applications to General Relativity. X, 220 pages. 2000.

Vol. m 62: J. Berger, J. Rubinstein, Connectivity and Superconductivity. XI, 246 pages. 2000.

Vol. m 63: R. J. Szabo, Ray Optics, Equivariant Cohomology and Localization of Path Integrals. XII, 315 pages. 2000.

Vol. m 64: I. G. Avramidi, Heat Kernel and Quantum Gravity. X, 143 pages. 2000.